LETTERS OF NOTE: OUTER SPACE

Letters of Note was born in 2009 with the launch of lettersofnote.com, a website celebrating old-fashioned correspondence that has since been visited over 100 million times. The first *Letters of Note* volume was published in October 2013, followed later that year by the first Letters Live, an event at which world-class performers delivered remarkable letters to a live audience.

Since then, these two siblings have grown side by side, with *Letters of Note* becoming an international phenomenon, and Letters Live shows being staged at iconic venues around the world, from London's Royal Albert Hall to the theatre at the Ace Hotel in Los Angeles.

You can find out more at lettersofnote.com and letterslive.com. And now you can also listen to the audio editions of the new series of *Letters of Note*, read by an extraordinary cast drawn from the wealth of talent that regularly takes part in the acclaimed Letters Live shows.

Letters of Note

OUTER SPACE

COMPILED BY

Shaun Usher

PENGUIN BOOKS

For the stars

PENGUIN BOOKS
An imprint of Penguin Random House LLC
penguinrandomhouse.com

First published in Great Britain as *Letters of Note: Space*
by Canongate Books Ltd 2021
Published in Penguin Books 2021

LIBRARY OF CONGRESS CATALOGING-IN-PUBLICATION DATA
Names: Usher, Shaun, 1978– compiler.
Title: Letters of note. Outer space / compiled by Shaun Usher.
Description: [New York] : Penguin Books, 2021. |
Series: Letters of note.
Identifiers: LCCN 2021021574 (print) | LCCN 2021021575 (ebook) |
ISBN 9780143134695 (paperback) | ISBN 9780525506508 (ebook)
Subjects: LCSH: Outer space. | Space sciences—Social aspects.
Classification: LCC QB500.25 .L48 2021 (print) |
LCC QB500.25 (ebook) | DDC 500.5—dc23
LC record available at https://lccn.loc.gov/2021021574
LC ebook record available at https://lccn.loc.gov/2021021575

Printed in the United States of America
1st Printing

Set in Joanna MT

CONTENTS

A letter is a time bomb, a message in a bottle, a spell, a cry for help, a story, an expression of concern, a ladle of love, a way to connect through words. This simple and brilliantly democratic art form remains a potent means of communication and, regardless of whatever technological revolution we are in the middle of, the letter lives and, like literature, it always will.

INTRODUCTION

Welcome to *Outer Space*.

As your eyes calmly flit from word to word on this page, the planet on which I presume you live is spinning on its axis at a speed of approximately 1,000 miles per hour, while racing around the Sun at 18.5 miles per second. At the same time, our solar system is also on the move, and not sluggishly either, hurtling around our galaxy, the Milky Way, at 140 miles per second. And do not think for one of those seconds that the Milky Way is stationary: it, too, is restless, racing towards the Andromeda Galaxy at 70 miles per second, the two destined to one day collide and cause humankind (if humankind has somehow managed to survive the mess we find ourselves in) all manner of problems. Put bluntly, we are part of something unimaginably huge, and all of it is moving at speed.

With that in mind, the small but perfectly formed book in your hands may feel incredibly quaint. And yet, in a way it is vast. *Letters of Note: Outer Space* is a collection of thirty letters on a subject grander than us all: letters written by astronauts, cosmonauts, astronomers, engineers, politicians, parents and children. They are letters of hope, awe, warning, complaint, remorse and fear. You will read a letter from a proud father to a space-bound son on the eve of his journey, and a farewell letter from a cosmonaut to his wife and daughters to be read should he never return from space. Particularly moving is a hopeful letter from an African-American girl desperate to reach the stars and the letter from a highly qualified pilot to the US President in which she pleads for the chance to join her male counterparts high above Earth.

As well as leaving me feeling minuscule, physically, researching this book also revealed my ignorance on many matters pertaining to space. I had no idea, for example, that Pluto was half the width of the United States, or that NASA employees so carefully considered the first words to be uttered by Neil Armstrong as he stepped onto the Moon. I was certainly unaware that the man who gifted us with the telephone believed so firmly that Mars was populated by extraterrestrials.

What did not surprise me, however, was that so many people have written fascinating letters about a place visited by so few of us. A destination so close yet so incredibly far. One can only imagine – and most of us have – what it must feel like to make the journey.

In 1997, twenty-eight years after walking on the Moon, Buzz Aldrin wrote the following to a professor about the experience:

'Someday in the future as people are mulling over their vacation plans, I hope they'll choose to fly into space. It's the trip of a lifetime.'

I hope so too.

Shaun Usher
2020

The Letters

THEY BLAZED A PATH FOR THE NEXT GENERATION

Akosua Haynes to Margot Lee Shetterly

2018

Published in 2016, Margot Lee Shetterly's Hidden Figures tells the true story of Dorothy Vaughan, Katherine Johnson and Mary Jackson, three pioneering African-American women who worked as 'human computers' at NASA in the 1960s. It was thanks to their pivotal calculations that celebrated astronauts such as Neil Armstrong and John Glenn could enter space, and yet for a long time their work remained uncelebrated. Shetterly's retelling of their remarkable story became a bestseller and an award-winning movie adaptation followed. In 2018, this letter was written to Shetterly by Akosua Haynes, a ten-year-old African-American girl with aspirations to become an astronaut one day.

THE LETTER

Dear Margot Lee Shetterly,

On August 21, 2017 I felt so lucky because it was the first day of school, and my friends were in class while I was watching the solar eclipse in Carbondale, Illinois. When the moon had completely covered the sun, I looked up and wondered how Katherine Johnson felt when she helped John Glenn orbit the earth. Reading your book "Hidden Figures" made me more excited about becoming a NASA astronaut, but it also made me question my career choice. It scared me when I read that a fireball entered into a spaceship killing all three astronauts inside. Becoming an astronaut had been my dream, I met Mae Jemison when I was four and have dressed up as an astronaut for at least four Halloweens, but I didn't want to die in a ball of flames.

I finished your book on the train ride back from Carbondale just five days before my "Hidden Figures" themed birthday party. I made up a rule, and told my friends that if they wanted to come they had to read at least two thirds of your book so that we could have an interesting discussion. I asked everyone to share their favorite passage. When it was my turn everyone read my selection, on page 217, aloud. Learning that John Glenn trusted

Katherine Johnson with his life, because of her superior math skills, motivated me to take my own math homework more seriously. I love math but some of my friends don't. I wanted them to read your book to see the magic in math and how useful it can be. Right before my party I looked up the definition for analytic geometry because Katherine used it to calculate the trajectory of John Glenn's Mercury capsule—useful magic!

Although John Glenn respected Katherine Johnson, they lived in two different worlds. When I read about the discrimination that Katherine and the computers had to put up with (people not trusting them and separate bathrooms), it made me think what it would have been like to live in the Jim Crow time period. I asked myself if I would have been able to work so well under pressure. I felt proud of Ms. Johnson.

There are many more opportunities for African Americans today because of what Katherine Johnson and the other computers accomplished. They blazed a path for the next generation. My friends thanked me for choosing your book to celebrate my birthday. I know that I can still be an astronaut, an astrophysicist, or have a space career on earth!

Sincerely,

Akosua Haynes

LETTER 02
I AM SO VERY ANXIOUS TO 'COME HOME'
Betty Trier Berry and Mount Wilson Observatory
21 January 1918

*In the early 1900s, decades before the women of
Hidden Figures played a pivotal role in the Space Race
– a time of heated competition between Cold War
rivals the USA and Russia (then the USSR) over space
exploration – the Mount Wilson Observatory hired
dozens of highly skilled women to do similar work. In
return for their invaluable efforts, they were offered
very little recognition and a paltry wage. As evidenced
by this exchange, despite her Masters degree in mathe-
matics and love of astronomy, Betty Trier Berry could
not afford to work as a 'human computer'. Instead,
she went on to become a celebrated attorney and the
first woman to work as a public defender in the
United States.*

THE LETTERS

Mrs. Betty Trier Berry
1929 No. Western Avenue,
Los Angeles, California

Dear Mrs. Berry,
We have a position in our Computing Division now vacant that I can offer you at a beginning salary of $825.00 per year, the appointment to take effect on February 1. An annual vacation of one month is granted to the staff, and there is no work on Saturday afternoons.

The amount offered is probably much less than you have been earning, but I am hoping that you will wish to try the work under these conditions. With your interest in astronomy, I am under the impression that you will not regret such an acceptance.

Very sincerely yours,
Superintendent Computing Division

January 23, 1918

My dear Mr. Seares,

Your kind communication of the 21st referring to a vacancy in your Computing Department at a $16 per week wage, reached me today; I trust you will pardon the intrusion on your time if I write you at some length on the matter.

I am dependent on my own efforts for my support; and desirous as I am of entering again the field of astronomical work, I am afraid of the ugly practical questions that would inevitably arise were I to limit my earning capacity to that rather pathetic amount. So far therefore as the position of which you write is concerned, I am regretfully obliged to say that it will be impossible for me to accept it.

It occurs to me, however, that you speak of my "interest in astronomy" as something quite apart from the work of the position you offer me. If I am correct in assuming from this that you have in mind for me, at a later time, some better position at a less mechanical, hence far more interesting, branch of the work, then it would seem that I am being given an opportunity to "try out," for which I am certainly grateful — but with all the risk of non-success resting upon me. As I say, I am scarcely in a position to assume this risk, and a salary which actually supports me during the probationary period would be obviously quite essential before I dared make a radical change of profession.

I am so very anxious to "come home" to the work I love and so confident that I can be of actual assistance to you in it, that I still venture to hope for an opportunity to join your staff.

Respectfully yours,

Betty T. Berry

* * *

January 30, 1918

Dear Mrs. Berry:

I am afraid that under existing conditions we can do nothing more than repeat the offer of my former letter; but from your reply I appreciate that that would be useless. I am sorry, for I had hoped that we might have you with us.

Very sincerely

LETTER 03
THE *VOYAGER* COSMIC GREETING CARD
Carl Sagan to Alan Lomax
6 June 1977

*Launched in 1977, the Voyager 1 space probe, as of
2020, was approximately 14,000,000,000 miles from
Earth. It reached interstellar space in 2012. Its sibling,
Voyager 2, was launched the same year. Aboard each
probe can be found a copy of the Voyager Golden
Record, a twelve-inch gold-plated copper disc on which
is stored hundreds of sounds and images indicative of
humankind – a time capsule for the attention of any
inquisitive extraterrestrials. The record's contents were
curated over the course of a year by a committee
headed by astronomer Carl Sagan. He wrote the
following letter in 1977 to esteemed musicologist Alan
Lomax, who had recently agreed to join the team.
Reprinted here too is a message from the US President
Jimmy Carter, also included on the Golden Record.*

THE LETTER

CORNELL UNIVERSITY
Center for Radiophysics and Space Research

June 6, 1977

Mr. Alan Lomax
215 West 98th Street
Apartment 12E
New York NY 10025

Dear Alan:
I am extremely pleased that you will be able to
lend us the benefit of your considerable experience
and expertise in ethnomusicology in the production
of the Voyager record.

Voyager 1 and Voyager 2 are unmanned deep space
probes which will be launched from Cape Canaveral
in August and September, 1977. Their mission is to
examine close-up the major planets, Jupiter, Saturn,
and Uranus, their 20 some odd moons, and the rings
of Saturn and Uranus. After these fly-by observations
are performed, the two spacecraft will be ejected
from the solar system, becoming mankind's third and
fourth interstellar space vehicles. The first two such
vehicles, Pioneers 10 and 11, were launched some

six years ago and contain a 6 x 9 inch gold anodized aluminium plaque on which is etched some simple scientific information about the location of the Earth and the solar system in the Milky Way Galaxy, and the moment in the ten billion year history of our Galaxy when the spacecraft was launched. There are also drawings of a man and woman. The plaques were a sort of message in a bottle, cast into the cosmic ocean, in case at some remote epoch in the future an extraterrestrial civilization were to come upon Pioneer 10 or 11 and wonder something about its origin.

Voyager permits us to continue on the Pioneer 10 and 11 experience, but in a much richer way. When NASA asked me to chair a committee to decide what should be the nature of the Voyager cosmic greeting card, it soon became clear that much more informa- tion could be conveyed in the same space on a metal mother of a phonograph record than on a plaque of the same size. Since this is the 100th anniversary of Edison's invention of the phonograph, a record seems particularly apt. NASA will be launching on each Voyager a bonded pair of copper mothers containing the equivalent of four sides of a 12-inch 33-1/3 rpm long playing record. One of these sides will contain digital scientific information – largely diagrams and pictures; a range of human voices, including some especially prepared at the United Nations and one special greeting by Kurt Waldheim, the U.N. Secretary General; and a selection of non-musical, non-vocal sounds of the Earth. The other three sides are devoted entirely to music – music representative of all of humanity and music which represents the best of humanity. We believe that public availability of a two-record album identical in content with the flight record will stimulate listeners to examine our civilization and culture and consider how we wish to be represented to the Cosmos. In addition, it may be for many people a first exposure to some of the diversity and quality of human music.

Under its protective cover the flight record will have a probable lifetime of a billion years. It is unlikely that many other artifacts of humanity will survive for so prodigious a period of time; it is clear, for example, that most of the present continents will be ground down and dissipated by then. Inclusion of the musical selections on the Voyager record ensures for them a kind of immortality which could not be achieved in any other way.

We would like you to be a member of the final musical selection committee and are very pleased that we will have access to some selections from your unique collection of ethnic music. In each case we will need a release from the copyright holder.

We feel we have an obligation in return for your important assistance to prominently acknowledge that assistance. We will be sure to include your name and affiliation in the draft of any NASA press release about the record. (We expect such releases around the time of launch in late August.) I would also like to invite you to prepare a few paragraphs for the commercial record liner or booklet, and – if there is a book on the Voyager record – to consider preparing all or part of a chapter on the rationale and significance of the particular selection of ethnic music which we will have made. I am enclosing a small token of my esteem.

With all good wishes,
Cordially,
Carl Sagan
Chairman
Voyager Record Committee

* * *

STATEMENT

This Voyager spacecraft was constructed by the United States of America. We are a community of 240 million human beings among the more than 4 billion who inhabit the planet Earth. We human beings are still divided into nation states, but these states are rapidly becoming a global civilization.

We cast this message into the cosmos. It is likely to survive a billion years into our future, when our civilization is profoundly altered and the surface of the Earth may be vastly changed. Of the 200 billion stars in the Milky Way galaxy, some − perhaps many − may have inhabited planets and space faring civilizations. If one such civilization intercepts Voyager and can understand these recorded contents, here is our message:

This is a present from a small distant world, a token of our sounds, our science, our images, our music, our thoughts, and our feelings. We are attempting to survive our time so we may live into yours. We hope some day, having solved the problems we face, to join a community of galactic civilizations. This record represents our hope and our determination and our goodwill in a vast and awesome universe.

[Signed]
Jimmy Carter
President of the United States of America
THE WHITE HOUSE
June 16, 1977

'WE HOPE SOME DAY,
HAVING SOLVED THE
PROBLEMS WE FACE,
TO JOIN A COMMUNITY
OF GALACTIC
CIVILIZATIONS.'
– President Jimmy Carter

LETTER 04
GO, JOHNNY, GO
Ann Druyan and Carl Sagan to Chuck Berry
15 October 1986

Nine years after Voyager 1 *and* 2 *left our planet,
astronomer Carl Sagan and writer, director Ann
Druyan wrote a letter to rock and roll legend Chuck
Berry, whose sixtieth birthday was approaching. He
was one of the artists selected to feature on the*
Voyager Golden Record, *the twelve-inch gold-plated
copper disc sent with the* Voyager *spacecraft, and
containing sounds and images selected to show the
diversity of life on Earth – the subject of Sagan's letter
to Alan Lomax (Letter 03).*

THE LETTER

Dear Chuck Berry,
When they tell you your music will live forever,
you can usually be sure they're exaggerating. But
Johnny B. Goode is on the Voyager interstellar
records attached to NASA's Voyager spacecraft—now
two billion miles from Earth and bound for the
stars. These records will last a billion years or more.

Happy 60th birthday, with our admiration for
the music you have given to this world . . .

Go, Johnny, go.

Ann Druyan
Carl Sagan

Cornell University,
Ithaca, New York
On behalf of the Voyager Interstellar Record
Committee

LETTER 05
TO A TOP SCIENTIST
Denis Cox and the Woomera Rocket Range
October 1957

*In 1957, it was announced that the Soviets had beaten
the US to create the first artificial Earth satellite, with
the successful launch of Sputnik 1. Following this news,
Australian schoolboy Denis Cox sent this urgent letter
to the Royal Australian Air Force's Rocket Range at
Woomera in an attempt to enter Australia into the
Space Race. Much to Denis's dismay, his letter,
addressed 'TO A TOP SCIENTIST', went unheeded –
until, fifty-two long years later, in 2009, Denis's original
letter and proposal for a rocket-ship design made the
news after being featured on the website of the
National Archives of Australia. As a result of the
coverage, he finally got a reply from the Australian
Department of Defence.*

THE LETTERS

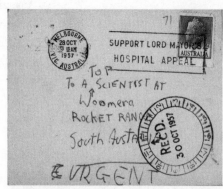

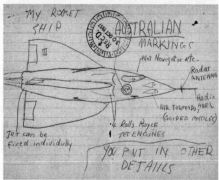

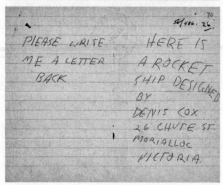

Australian Government
Department of Defence
Defence Science and
Technology Organisation

Mr Denis Cox

28/8/09

Dear Mr Cox,

I would like to thank you for your letter we received on 20th Oct. 1957 regarding the design of your rocketship. I apologise for the late response to your letters. You will appreciate that as you requested "A Top Scientist" that uses the "WOOMERA ROCKET RANGE" it took a little while for your letter to get to me and in addition, it took some time to provide due consideration to your ideas.

In any case, I have included a picture of our latest flight of a hypersonic vehicle under the HIFiRE Program, so that you may see that many of your designs have merit. The fins are a little smaller, and on our work, we haven't advanced sufficiently to put people on board, as you clearly indicated we should. Curiously enough though, people are still toying with the idea of combining rocket engines with turbines as suggested in your letter. These engines are now called Rocket Based Combined Cycle Engines and seem to work about as well as they did back in 1957! I am also quite interested in the shape of the fuselage, it actually shows a lot of merit!

I think that the most interesting statement you made in your letter was "YOU PUT IN OTHER DETAILS". You were clearly going to be an excellent Program Manager, by providing those that know best the freedom in the matter to get it right. Furthermore, you did have your priorities right as "AUSTRALIAN MARKINGS" are the most prominent feature of the design.

I remember as a boy designing rocket ships and planes at about the same time that you wrote your letter. I don't know why or how, but somehow I was lucky enough to get to a position where I now head a team that designs planes and engines that will soon fly at Mach 8, or around 9000km/hr. I am proud to tell you that these planes will have an "AUSTRALIAN MARKING" on them as you indicated they should have. My one hope is that we do a sufficiently good job that is worthy of the inspiration, dreams and hopes that you provided in your letter those many years ago.

Once again, thank you for your letter.

(Signed)
Allan Paul BSc PhD MEngSc
Research Leader Applied Hypersonics
Air Vehicles Division
DSTO-Brisbane

LETTER 06
VOYAGE FROM THE EARTH TO THE MOON
Frank Borman to Jean Jules-Verne
5 February 1969

*On Christmas Eve 1968, three days after launching
from Florida's Kennedy Space Center, Apollo 8 became
the first crewed spacecraft to orbit the Moon – a feat
they managed not just once but ten times during their
mission. On board were commander Frank Borman and
his crewmates Jim Lovell and Bill Anders, all of whom
took turns reading the first ten verses from the Book
of Genesis. Their recitals were beamed back to Earth
and broadcast to millions. Two months after this
historic flight, Borman wrote a letter to Jean Jules-
Verne, grandson of Jean Jules-Verne, the French novelist
who 104 years earlier had written of such a trip in his
eerily prescient book From the Earth to the Moon.*

THE LETTER

NATIONAL AERONAUTICS AND SPACE
ADMINISTRATION

February 5, 1969

Mr. Jean Jules Verne
Le Vieil Huba
83 – Toulon

Dear Sir:

As you undoubtedly know, generations of American
schoolboys – including Captain Lovell, Major
Anders and me – have been fascinated by the books
of your illustrious grandfather.

That his Voyage from the Earth to the Moon
foresaw both the fact of the magnificent adventure
which mankind has just experienced and even such
details as liftoff from Florida and splashdown in the
Pacific is more than an extraordinary coincidence: it
is a tribute to the genius of his vision. He not only
imagined what exploits were possible for man but
even how they might be accomplished, down to
the finest details.

Who can say how many of the world's space
scientists were inspired, consciously or unconsciously,

by their boyhood reading of the works of Jules Verne? What can be said is that behind every breakthrough in man's history there lies a dream. That dream, and the blueprint for its accomplishment, were first shown to us by your grandfather. In a very real sense, then, Jules Verne is one of the great pioneers of the space age.

Very sincerely yours,
Frank Borman

LETTER 07
E.T. HAS CHANGED TOMMY'S LIFE
Various to E.T. and Steven Spielberg
Circa 1982

Few movies have ever managed to affect an audience quite like E.T. the Extra-Terrestrial. Released in 1982, Steven Spielberg's genre-defining tale of friendship between boy and alien quickly broke box-office records, won multiple Academy Awards and captured countless hearts, all the while provoking people of all ages to look up and wonder about the possibility of life on planets beyond our own. Naturally, a cultural event so impactful also generated letters, and here are just three sent to Spielberg and E.T. himself (via his fan club).

THE LETTERS

Dear Mr. Spielberg,

I realize this letter must be one of thousands you have received since the release of E.T. And it's really not that extraordinarily different from most of those letters. I suppose I could quite truthfully rant and rave about what a marvelous movie E.T. is and what exceptional talent you have demonstrated in your direction of it, but I won't.

When I was a child growing up I dreamed of meeting life from other worlds. I can remember staring off into the heavens outside my window and hoping if I wished hard (and sincerely) enough that the inhabitants of the planetary system circling the only star I could see through my elm tree would come and visit me.

I would go out at dusk and string yards of brightly colored wires on my dilapidated wire mesh fence. Maybe by some miraculous freak of physics I could somehow send a message to my star. And They would come.

Now the wires are gone, and the fence shows no signs of the enormous weight of dreams it once bore. The elm tree that once blocked all stars but my own got Dutch elm disease and lets hundreds of stars shine through its branches. And I have grown old, keeping childhood dreams hidden and tucked away.

E.T. found that favorite fantasy and let me live it again. You did justice to the dream of a child.

THANK YOU, STEVEN SPIELBERG.

Rita Calm

September 14, 1982

* * *

Dear E.T.

I think you're cute and funny. Where do you come from? What are your friends like? What is your Space Ship like? How fast can your Space Ship fly? What is your home like? What language do you speak? How do you like earth? Maybe you can't answer all of these questions but just answer some of them, and please send me a small tollken [sic] of something.

Sincerely,

Mark

* * *

Dear E.T. Fan Club (and Mr. Spielberg),

I am Tommy's mother. I am writing this letter for

him as Tom has never really learned to write much more than his first name. Tom is 20 and autistic. That means he prefers his own strange world to the real one outside himself. Since he has always enjoyed movies filled with special effects, space-crafts and startling aliens, it was only natural for his parents to take turns waiting in the long lines for E.T. In the darkened theater, Tommy came out of himself. He screamed—he clapped—he laughed . . . and then—yes—Tommy cried. Real tears. Autistics do not weep—not for themselves or any others. But Tommy wept and Tommy talked—nonstop—about E.T . . . Tom has seen E.T. three times now and is prone to touching fingers with others and solemnly repeating, "Ouch."

E.T. has changed Tommy's life. It has made him relate to something beyond himself. It's as though Tommy has also been an alien life-form and trying to find his way home—just like E.T.

Totally yours,
Ann Andonian

LETTER 08
IF YOU ARE TO BE, BE THE FIRST
Yuri Gagarin to his family
10 April 1961

Yuri Gagarin was born in 1934 in the Russian village of
Klushino, to Alexey Ivanovich Gagarin and Anna
Timofeyevna Gagarina – a carpenter and farmer,
respectively. Aged twenty-three, Gagarin met and fell in
love with a nursing student from Orenburg named
Valentina Goryacheva. Within months they were
married, going on to have two daughters, Yelena and
Galina. On 12 April 1961, Gagarin waved goodbye to
his young family, stepped into the Vostok 1 spacecraft
and became the first man to go into space. He
returned safely; however, he had planned for a
different outcome, and two days before launch he
wrote this letter to his wife and children, to be opened
in the event of his death.

THE LETTER

My dear beloved Valechka, Lenochka, and Galochka!

I decided to write you a few lines to share with you the joy and happiness that befell me today. Today the government commission decided to send me to the first spaceflight. I am so glad, dear Valiusha, and I want you to share this joy with me. An ordinary man, I have been trusted with an important national mission—to pave the first road into space! [. . .]

I fully trust the technology. It should not fail. But it sometimes happens that a man falls and breaks his neck with no reason at all. Something may happen here too. I do not believe it will happen. But if it does, I ask all of you and especially you, Valiusha,—do not be overcome with grief. Such is life [. . .] Please take care of our girls and love them just like I do. Please raise them as [. . .] true human beings who are not afraid of the challenges of life. Raise them as people who will deserve to live a new communist society.

This letter is coming out a bit too gloomy. I do not believe in this [bad] outcome. I hope you will never see this letter, and I will feel ashamed for this momentary weakness. But if something happens, you must know the whole truth

When I was a child, I once read the words of Valerii Chkalov, "If you are to be, be the first." So I try, and I will to the end. Valechka, I wish to dedicate this flight to the people of the new communist society, which we are already entering, to our great Motherland, and to our science

I hope in a few days we will be together again, and we will be happy. Valechka, please, do not forget my parents, and if you have an opportunity, help them somehow. Give them my warm greetings, and let them forgive me for not telling them about this, for they were not supposed to know. This is it, I think.

Good bye, my dears. I hug you tightly and kiss you.

Greetings

Your daddy and Yura.

'AN ORDINARY MAN,
I HAVE BEEN TRUSTED
WITH AN IMPORTANT
NATIONAL MISSION.'

— Yuri Gagarin

LETTER 09
SOCIALISM IS THE BEST LAUNCHING PAD
FOR SPACE FLIGHTS

Soviet cosmonauts to Leonid Ilyich Brezhnev
22 October 1965

*Following Yuri Gagarin's historic journey into space in
1961, he quickly became a global celebrity. He was
twenty-seven, handsome and charismatic – a heady
combination – and travelled the world as a spokesman
for his country's space programme and, by extension,
political system. However, over the next few years, the
Space Race gathered pace, and by 1965 it was clear to
Gagarin and his fellow cosmonauts that the United
States had pulled far ahead of the Soviet Union. This
critical letter to Soviet leader Leonid Ilyich Brezhnev
was their plea for change. Gagarin died in 1968, when
the jet in which he was flying crashed in the Russian
town of Kirzhach. A year after his death, NASA landed
men on the Moon.*

THE LETTER

Central Committee of the Communist
Party of the Soviet Union
Comrade L. I. Brezhnev

Dear Leonid Il'ich!

We are writing to you to raise certain issues, which we consider very important for our state and for us.

Soviet achievements in space exploration are well known, and there is no need to list all of our victories here. These victories have been achieved and will remain in history to be the pride of our nation forever. The people, the Party and our leaders have always appropriately connected our achievements in space with our achievements in the construction of socialism. "Socialism is the best launching pad for space flights." This catchphrase circled the entire world. Soviet people said these words with pride, the peoples of the socialist countries believed it was true, and hundreds of millions of people abroad learned the ABC of communism through our achievements in space. Such it was. We, cosmonauts, travelled abroad many times; a thousand times we witnessed how warmly multi-million crowds in various countries greeted Soviet achievements in space.

In the past year, however, the situation has changed. The USA have not only caught up with us, but even surpassed us in certain areas. The flights of space vehicles Ranger-7, Ranger-8, Mariner-4, Gemini-5 and others are serious achievements of American scientists.

This lagging behind of our homeland in space exploration is especially objectionable to us, cosmonauts, but it also damages the prestige of the Soviet Union and has a negative effect on the defence efforts of the countries from the socialist camp.

Why is the Soviet Union losing its leading position in space research? A common answer to this question answer [sic] is as follows: the USA have developed a very wide front of research in space; they allocate enormous funds for space research. In the past 5 years they spent more than 20 billion dollars, and in 1965 alone 7 billion dollars. This answer is basically correct. It is well known that the USA spend on space exploration much more than does the USSR.

But the matter is not only funding. The Soviet Union also allocates significant funds for space exploration. Unfortunately, in our country there are many defects in planning, organisation, and management of this work. How can one speak about serious planning of space research if we do not have any plan for cosmonauts' flights? The month of October is coming to an end, there is a little time left before the end of the year 1965, but no one in Soviet Union knows whether there will be a manned space flight this year, what will be the task for that flight, and what duration. The same situation was characteristic of all the previous flights of the ship-satellites Vostok and Voskhod. This creates totally abnormal conditions during cosmonauts' preparation for flight and precludes the possibility of preparing crews for flight without hassle ahead of time.

We know that in this country there are plans for developing space technology, we know decisions of the Central Committee of the CPSU and the government that include specific deadlines for the construction of spacecrafts. But we know also that many of these decisions are not being implemented at all, and most are being carried out with huge delays.

Manned space flights are becoming more and more complex and prolonged. The preparation of such flights takes a lot of time, requires special equipment, training spacecraft and simulators, which are now being created with huge delay and with primitive methods. To put it briefly, we need a national plan of manned space flights which would include the flight task, the date, the composition of the crew, the duration of the flight, the deadline for the preparation of a spacecraft and a simulator, and many other important issues of flight preparation.

Up to now manned space flights have been carried out according to the plans of the USSR Academy of Sciences, while the direct management and technical support have been organised by representatives of the industry and the USSR Ministry of Defence. Items of military significance have been present in flight programmes only to some degree, which can be explained by the fact that within the Ministry of Defence there is no organisation that would unify the whole complex of questions of space exploration. Everybody is involved in space affairs – the Missile Forces, the Air Force, the Air Defence, the Navy and other organisations. Such scattering of efforts and resources in space exploration interferes with work; a lot of time is spent on coordination of plans and decisions, and these decisions often reflect narrow departmental interests. The existing situation with the organisation of space research contradicts the spirit of the decisions of the September Plenum of the Central Committee of the CPSU, and it must be changed.

In 1964 the chief of the Joint Staff, the Marshal of the Soviet Union Biriuzov created a special commission. This commission studied in detail the organisation of work on space exploration and came to the conclusion that it was necessary to unify all space affairs under the command of the Air Force. The Marshal of Soviet Union S.S. Biriuzov, the General of the Army A.A. Epishev, and the Marshal of the Soviet Union A.A. Grechko supported this proposal. But after the tragic death of the Marshal of the Soviet Union Biriuzov this reasonable proposal was discarded and the Central Administration for Space Exploration (TsUKOS) was organised under the Missile Forces. The creation of this organisation changed nothing, however. The narrow departmental approach, the scattering of resources, and the lack of coordination have persisted.

The Air Force leadership and we, cosmonauts, repeatedly addressed the Joint Staff, to the Minister of Defense, and to the Military-Industrial Commission with specific proposals on the construction of and the equipment for spacecrafts that would be capable of carrying out military tasks. As a rule, our proposals were not supported by the Missile Forces leadership. We received such replies as: "Vostok spacecraft do not have any military value, and it is inexpedient to order their construction" and "We will not order Voskhod spacecraft, for there are no funds".

In 1961 we had two Vostok spacecraft.

In 1962 we had two Vostok spacecraft.

In 1963 we had two Vostok spacecraft.

In 1964 we had one Voskhod spacecraft.

In 1965 we had one Voskhod spacecraft.

In 1965 the Americans launched three Gemini spacecraft, and they are planning to launch two more before the end of the year.

Why have not enough ships been built for our cosmonauts' flights? In any case, not because of the lack of funding. It happened because the leadership of the Missile Forces has more trust in automatic satellites, and it underestimates the role of human beings in space research. It is a shame that in our country, which was the first to sent man into outer space, for four years the question has been debated whether man is needed on board a military spacecraft. In America this question has been resolved firmly and conclusively in favour of man. In this country, many still argue for automata. Only these considerations can explain why we build only 1–2 piloted ships in the same period as 30–40 automatic satellites are being produced. Many automatic satellites cost much more than a piloted ship, and many of them never reach their destination. The Vostok and the Voskhod piloted spacecraft have carried out a full programme of scientific research and at the same time have produced a huge political effect for this country.

We do not intend to belittle the value of automatic spacecraft. But an infatuation with them would be, at the very least, harmful. Using the Vostok and the Voskhod spacecraft, it would have been possible to carry out a large complex of very important military research and to extend the duration of flights to 10–12 days. But we have no ships, nothing on which we could fly, nothing on which we could carry out a programme of space research.

Besides what is stated above, there are also other defects in the organisation of our flights – defects which we cannot remedy by ourselves. In our country there is no unified centre for space flight control. During the flight every spacecraft has no communication with the command station in between the sixth and the thirteenth turn circuits of the day. At the testing range, there are bad conditions for training and resting of cosmonauts.

We also have other questions awaiting a resolution. Many questions could be resolved without appealing to the Central Committee of the CPSU. We repeatedly wrote to the Minister of Defence about these questions. We are aware of the petitions from the Air Force leadership to the Ministry of Defence and the government, but these petitions largely did not fulfill their purpose. Many times we met with the Minister of Defence, but unfortunately those were not business meetings. And today we have no confidence that the issues we raise can be resolved at the Ministry of Defence.

Dear Leonid Il'ich! We know how busy you are and nevertheless we ask you to familiarise yourself with our space affairs and needs.

The 50th anniversary of the Great October [Revolution] is approaching. We would like very much to achieve new big victories in space by the time of this great holiday.

We are deeply convinced that resolving the issue of unifying all military space affairs under the command of the Air Force, the thoughtful planning of space research, and the construction of space-craft that would solve the problem of military application of piloted spacecraft would appreciably strengthen the defensive power of our homeland.

Pilots-cosmonauts of the USSR

Yu. Gagarin

A. Leonov

P. Belyaev

G. Titov

A. Nikolaev

V. Bykovsky

October 22, 1965

LETTER 10
THE RESULT WOULD BE A CATASTROPHE
Roger Boisjoly to R. K. Lund
31 July 1985

In a tragedy witnessed by millions, Space Shuttle
Challenger *broke apart over the coast of Florida just
seventy-three seconds after launch on 28 January 1986,
ending the lives of all seven of its crew members. A
subsequent investigation determined that the accident
had been caused by the failure of an O-ring – essentially a rubber seal on one of the shuttle's solid rocket
boosters – brought on, in part, by extremely cold
weather around the time of launch. But the development wasn't a shock to all involved. Six months prior
to the launch, this memo was sent by Roger Boisjoly,
an engineer working at Morton Thiokol, the manufac-
turers of the solid rocket boosters, to the company's
vice-president. In it, he predicted the problem and
warned of a potential disaster. Boisjoly's warning went
unheeded. He later attempted to halt the launch,
unsuccessfully.*

THE LETTER

31 July 1985

TO: R. K. Lund Vice President, Engineering
FROM: R. M. Boisjoly, Applied Mechanics - Ext. 3525
SUBJECT: SRM O-Ring Erosion/Potential Failure
Criticality

This letter is written to insure that management
is fully aware of the seriousness of the current
O-Ring erosion problem in the SRM joints from
an engineering standpoint.

The mistakenly accepted position on the joint
problem was to fly without fear of failure and to
run a series of design evaluations which would
ultimately lead to a solution or at least a significant reduction of the erosion problem. This
position is now drastically changed as a result of
the SRM 16A nozzle joint erosion which eroded
a secondary O-Ring with the primary O-Ring
never sealing.

If the same scenario should occur in a field joint
(and it could), then it is a jump ball as to the
success or failure of the joint because the secondary
O-Ring cannot respond to the clevis opening rate
and may not be capable of pressurization. The result

would be a catastrophe of the highest order – loss of human life.

An unofficial team [a memo defining the team and its purpose was never published] with leader was formed on 19 July 1985 and was tasked with solving the problem for both the short and long term. This unofficial team is essentially nonexistent at this time. In my opinion, the team must be officially given the responsibility and the authority to execute the work that needs to be done on a non-interference basis (full time assignment until completed.)

It is my honest and very real fear that if we do not take immediate action to dedicate a team to solve the problem with the field joint having the number one priority, then we stand in jeopardy of losing a flight along with all the launch pad facilities.

[Signed]
R. M. Boisjoly

Concurred by:
[Signed]
J. R. Kapp, Manager
Applied Mechanics

LETTER 11

AND TEARS DON'T FLOW THE SAME IN SPACE

Frank L. Culbertson Jr to the people of Earth

12 September 2001

On the morning of 11 September 2001 the world looked on in horror as four commercial jets were hijacked by terrorists and used as weapons to kill almost 3,000 people on US soil. Two of those planes were flown into the iconic Twin Towers of New York City's World Trade Center, ultimately bringing them both to the ground. There was just one American not on Earth that morning: Frank Culbertson, a fifty-two-year-old NASA astronaut. At the time, he was 250 miles from his home planet, working as commander of the International Space Station. A day after the attacks, he wrote home.

THE LETTER

I haven't written very much about specifics of this mission during the month I've been here, mainly for two reasons: the first being that there has been very little time to do that kind of writing, and secondly because I'm not sure how comfortable I am sharing thoughts I share with family and friends with the rest of the world.

Well, obviously the world changed today. What I say or do is very minor compared to the significance of what happened to our country today when it was attacked by . . . by whom? Terrorists is all we know, I guess. Hard to know at whom to direct our anger and fear . . .

I had just finished a number of tasks this morning, the most time-consuming being the physical exams of all crew members. In a private conversation following that, the flight surgeon told me they were having a very bad day on the ground. I had no idea.

[. . .]

I know so many people in Washington, so many people who travel to DC and NYC, so many who are pilots, that I felt sure I would receive at least a

47

few pieces of bad news over the next few days. I got the first one today when I learned that the Captain of the American Airlines jet that hit the Pentagon was Chic Burlingame, a classmate of mine. I met Chic during plebe summer when we were in the D&B [the Drum and Bugle corps] together, and we had lots of classes together. I can't imagine what he must of gone through, and now I hear that he may have risen further than we can even think of by possibly preventing his plane from being the one to attack the White House. What a terrible loss, but I'm sure Chic was fighting bravely to the end. And tears don't flow the same in space . . .

It's difficult to describe how it feels to be the only American completely off the planet at a time such as this. The feeling that I should be there with all of you, dealing with this, helping in some way, is overwhelming. I know that we are on the threshold (or beyond) of a terrible shift in the history of the world. Many things will never be the same again after September 11, 2001. Not just for the thousands and thousands of people directly affected by these horrendous acts of terrorism, but probably for all of us. We will find ourselves feeling differently about dozens of things, including prob- ably space exploration, unfortunately.

It's horrible to see smoke pouring from wounds in your own country from such a fantastic vantage point. The dichotomy of being on a spacecraft dedicated to improving life on earth and watching life being destroyed by such willful, terrible acts is jolting to the psyche, no matter who you are. And the knowledge that everything will be different than when we launched by the time we land is a little disconcerting. I have confidence in our country and in our leadership that we will do everything possible to better defend her and our families, and to bring justice for what has been done. I have confidence that the good people at NASA will do everything necessary to continue our mission safely and return us safely at the right time. And I miss all of you very much. I can't be there with you in person, and we have a long way to go to complete our mission, but be certain that my heart is with you, and know you are in my prayers.

Humbly,

Frank

LETTER 12
KNOWLEDGE BEGETS KNOWLEDGE
Mary Lou Reitler and John F. Kennedy
19 January 1962

US President John F. Kennedy made an historic speech before Congress in May 1961 – one month after Soviet cosmonaut Yuri Gagarin became the first person to be launched into outer space and orbit Earth – proposing 'landing a man on the Moon and returning him safely to the Earth' before the decade's end. Eight years and billions of dollars later, Kennedy's goal was achieved when Neil Armstrong and Buzz Aldrin set foot on the lunar surface. Early in 1962, eight months after Kennedy's call to action, a teenager named Mary Lou Reitler wrote to him and questioned the need for humans to explore space. A response subsequently arrived from Myer Feldman, Deputy Special Counsel to the President.

THE LETTERS

January 19, 1962
R.F.D. #1
Delton, Michigan

Dear President Kennedy,
I am thirteen years old and I'm in the eighth grade.
Please don't throw my letter away until you've read
what I have to say. Would you please answer me
this one question? When God created the world, He
sent man out to make a living with the tools He
provided them with. They had to make their living
on their own with what little they had. If He had
wanted us to orbit the earth, reach the moon, or
live on any of the planets, I believe He would have
put us up there Himself or He would have given us
missiles etc. to get there. While our country is
spending billions of dollars on things we can get
along without, while many refugees and other
people are starving or trying to make a decent
living to support their families. I think it is all just
a waste of time and money when many talents
could be put to better use in many ways, such as
making our world a better place to live in. We
don't really need space vehicles. I think our country
should try to look out more for the welfare of its

people so that we can be proud of the world we live in. At school they tell us that we study science so that we can make our world a better place to live in. But I don't think we need outer space travel to prove or further the development of the idea. Now that you have heard what I have to say will you please write me in answer to my question?

Sincerely,

Mary Lou Reitler

* * *

March 29, 1962

Dear Mary Lou Reitler

The President has asked me to reply to your letter asking why the United States expends so much time and energy in exploring space, and suggesting that God would have provided man with the necessary space implements had he wanted man to explore space.

A significant feature of our society is the right of each individual to determine the nature of God's intent in accordance with his own conscience. I would not, therefore, presume to suggest how you should resolve the issue you pose in your letter. Yet it would appear that among the most common characteristics of man is a desire to impose change on nature in order to mollify the hardships of life. This, combined with an endowed natural intelligence and curiosity, has allowed man to progress through increased knowledge from the most primitive past when the only tools utilized were those rocks and sticks found lying on the ground, to the present day, when disease-controlling drugs, efficient food production, and labor-saving machinery have combined to permit man, if he wishes, to pursue a far richer and more humane life.

It is impossible to determine in advance, moreover, those benefits which will eventually result from a given advance in human knowledge. History is replete with examples of man pursuing knowledge with no expectation of any practical use, which later serve as the basis for developments making significant contributions to mankind. Janssen's work on lenses, without his realizing it, provided the breakthrough required to understand and control disease-causing microbes; also, Hertz predicted that his academic experimentation with electromagnetic waves would have no practical or useful result, but he had in fact helped to lay the groundwork for the modern electronic industry.

Astronaut John H. Glenn briefly summed this up recently when he explained his views on the importance of space research before a joint session of Congress. He in part said, "But exploration and the pursuit of knowledge have always paid dividends in the long run – usually far greater than anything expected at the outset . . . Any major effort such as this results in research by so many different specialties that it is hard to even envision the benefits that will accrue in many fields. Knowledge begets knowledge. The more I see, the more impressed I am – not with how much we know – but with how tremendous the areas are that are as yet unexplored."

Thank you very much for advising the President of your views on this important matter. It is encouraging to find someone of your age showing such interest and concern in public affairs.

Sincerely,

Myer Feldman

Deputy Special Counsel

to the President

'IT IS IMPOSSIBLE TO DETERMINE IN ADVANCE, MOREOVER, THOSE BENEFITS WHICH WILL EVENTUALLY RESULT FROM A GIVEN ADVANCE IN HUMAN KNOWLEDGE.'

— Myer Feldman

LETTER 13
MAN IN SPACE
Alan Shepard to his parents
29 January 1959

On 5 May 1961, Alan Shepard became the second person – and first American – to enter space. This letter was written by Shepard to his parents two years earlier and marks the very first announcement of his plans to volunteer for the 'Man in Space' programme. In fact, the very same day, hours after penning these words, Shepard and a group of other specially selected pilots travelled to Washington, where they were briefed about Project Mercury by NASA for the first time. He was chosen as one of the 'Mercury Seven' just over two months later, at which point he began the training that would ultimately see him take the Freedom 7 *spacecraft to an altitude of 187 kilometres. A decade after that, Shepard became the fifth person to walk on the Moon.*

THE LETTER

Dear Mother and Daddy—
Thanks so much for your recent note, Daddy. I
appreciate also you sharing the commission on my
insurance premium.

We are enjoying our time here very much and
like the new addition to our house. It makes it
much more livable. Room for guests—so come on
down!

Present plans are to be up for Ann's wedding in
April. No details as yet but will keep you posted.

I am driving to Washington this afternoon for
a briefing and for consideration in the "Man in
Space" program. I am letting you know right away
since I am not sure how much publicity or press
releases will be involved. Basically, about 100 of
the country's top pilots have been selected to go
to Washington to be briefed on the plans for
putting a man in space some time during 1961.
We are to be given a chance to volunteer for or
reject the opportunity after the briefing. Thereafter,
all volunteers will go through a rigorous elimina-
tion process until a handful are selected. The
entire program of space travel is a fascinating

subject and I'm very pleased to be associated with it!

I assure you that I will analyze the entire picture based upon my past flight experience. I intend to do it very carefully of course—and will most certainly volunteer for it. There is no reason for expression of fear but merely gratitude to be considered for this very important contribution to science and the country. Will keep you posted.

Please make no announcements or statements at this time should the occasion arise or even if it doesn't arise!

My love to you both—
Alan

'THERE IS NO REASON
FOR EXPRESSION OF
FEAR BUT MERELY
GRATITUDE TO BE
CONSIDERED FOR THIS
VERY IMPORTANT
CONTRIBUTION TO
SCIENCE AND THE
COUNTRY.'

— Alan Shepard

LETTER 14
THE SUN & THE COMET WAS TO HAVE A FIGHT

Nellie Copeland to Dr William R. Kubinec

February 1985

Halley's Comet is arguably the most famous of all comets. Five miles wide and ten miles long, it is named after eighteenth-century astronomer Edmund Halley, who deduced that its periodic appearances – it can be spotted from Earth once every seventy-five to seventy-six years – were those of a returning body. The first recorded sighting of Halley's Comet occurred in 239 BC, and the most recent appearance was in 1986. In July of the previous year, in anticipation of the comet's return, Dr William R. Kubinec, chairman of the Department of Physics at the College of Charleston, West Virginia, published a request for recollections of its 1910 appearance. This eyewitness account soon arrived, written by an elderly lady named Nellie Copeland.

THE LETTER

Maryville, Tennessee
Feb – 6 – 1985

Dear Sir

I just read in my Paper the Maryville Times about
the piece about Halleys Comet and it interested me
for I remember seeing it. I was borned May 15 –
1954 which makes me 80 years young. I remember
it was almost Sun-down and the word got around
the Sun & the Comet was to have a fight. The
neighbors all gathered on a hillside to watch.

My grand-mother was a very nervous old lady.
So it was just before Sun-down then Halleys Comet
past the sun which was real red & the tail of the
Comit was real wide and long and the Sun seem to
tremble & shake. It was a wonderful sight.

I'm now looking forward to living to seeing it
again when it comes back.

Some people thought the world was coming to
an end.

Hoping you see it next time for it was a
wonderful sight.

Yours truly,

Mrs. Nellie Copeland

Good luck to you.

LETTER 15
MY SISTER SAYS I AM AN ALIEN
Jack Davis and NASA
3 August 2017

As we humans expand our footprint and ever so slowly explore wider expanses of this unthinkably vast universe, it is vital that we do so with great care to minimise the risk of interplanetary contamination – both inbound (space to Earth) and outbound (Earth to space). Since 1967, space missions have been bound by the United Nations' Outer Space Treaty to avoid such harmful contamination. As of 2020, it is joined by 110 countries. In 2017, NASA publicly announced that it was looking for a new 'Planetary Protection Officer'. Unsurprisingly, countless applications arrived, including a letter from nine-year-old Jack Davis.

THE LETTERS

Dear NASA,
My name is Jack Davis and I would like to apply
for the planetary protection officer job. I may
be nine but I think I would be fit for the job.
One of the reasons is my sister says I am an alien.
Also, I have seen almost all the space and alien
movies I can see. I have also seen the show
Marvel Agents of Shield and hope to see the
movie Men in Black. I am great at vidieo games.
I am young, so I can learn to think like an
Alien.

   Sincerely,
   Jack Davis
   Guardian of the Galaxy
   Fourth Grade

* * *

NASA

Dear Jack,
I hear you are a "Guardian of the Galaxy" and that

you're interested in being a NASA Planetary Protection Officer. That's great!

Our Planetary Protection Officer position is really cool and is very important work. It's about protecting Earth from tiny microbes when we bring back samples from the Moon, asteroids and Mars. It's also about protecting other planets and moons from our germs as we responsibly explore the Solar System.

We are always looking for bright future scientists and engineers to help us, so I hope you will study hard and do well in school. We hope to see you here at NASA one of these days!

Sincerely,

Dr. James L. Green

Director, Planetary Science Division

'I AM YOUNG, SO I CAN LEARN TO THINK LIKE AN ALIEN.'

— Jack Davis

LETTER 16
I CAN BE PATIENT NO LONGER
Jerrie Cobb to John F. Kennedy
13 March 1963

*Born in Oklahoma in 1931 to Lt. Col. William H. Cobb
and Helena Butler Stone Cobb, Geraldyn 'Jerrie' Cobb
was twelve years old when she first took the controls
of her father's 1936 Waco aircraft. From that day, she
never looked down. At sixteen, she had a private
pilot's licence; at eighteen, she was a certified ground
instructor; soon she was setting various world aviation
records for speed, distance and altitude. In 1960, she
was the first of thirteen women (the 'Mercury 13')
selected by NASA to undergo intensive training that
would determine whether a woman could become an
astronaut. Despite passing every test, she remained
grounded and soon the programme was cancelled. In
1963, having brought the issue before Congress to no
avail, Cobb wrote this letter to President John F.
Kennedy, and pleaded to be sent into space immediately. To her frustration, it would be another twenty
years until an American woman, Sally Ride, left Earth's
orbit. Three months after this letter was written,
cosmonaut Valentina Vladimirovna Tereshkova became
the first woman in space.*

THE LETTER

March 13, 1963

The President
The White House
Washington, D. C.

Dear Mr. President:
It is difficult to write this letter knowing it will be
read by your secretaries and assistants and the
chances are slim that it will get through to you. I
feel compelled to do so anyway, in the faith that
the matter will in some way be brought to your
attention.

Some of your staff are acquainted with my
efforts to get the United States to put the first
woman in space. For three years I have been
working for this (including passing the three
phases of astronaut testing). I have discussed this
matter with Vice President Johnson and Dr. Welsh
of your Space Council as well as many of our
country's top space scientists. The reaction has been
one of general acceptance: "why don't we do it
now, before Russia?"; "the scientific reasons more
than justify the cost"; "what are we waiting for?"
are typical of the response. That is, with all except

the top echelon of NASA. James Webb appointed
me a consultant to NASA over two years ago but
never used my services.

I have not wanted to bother you with this
matter but I can be patient no longer. It is a fact
that the American people want the United States to
put the first woman in space. While NASA refuses,
the Soviet Union openly boasts that they will
capture this next important scientific first in space
by putting their lady cosmonaut up this year. We
could have accomplished this scientific feat last
year, and even now, could still beat the U.S.S.R. if
you would make the decision. It need not even be
a long orbital shot, or interfere with the current
space programs; on a rush basis a sub-orbital shot
would suffice or a X-15 flight to a 50 mile altitude.
Any aerospace doctor or scientist will tell you the
scientific data obtained from such an experiment
would be of lasting benefit.

Enclosed is a file of my correspondence with
NASA, a scrapbook and several clippings. I have
worked, studied and prayed for this over three
years now and could not give up without one last,
final plea to the commander-in-chief. Forgive me
for taking up your time but I still believe the
matter is of utmost importance, worthy of your

serious consideration; and may the Lord guide you in your decision.

I have the honor to remain,
Your most obedient servant,
[Signed]
Miss Jerrie Cobb

LETTER 17
I AM SO PROUD OF YOU, OUR SOVIET GIRL

Valentina Vladimirovna Zorkina to Valentina
Vladimirovna Tereshkova
20 June 1963

*On 16 June 1963, two years after Yuri Gagarin became
the first man to journey into space, twenty-six-year-old
Valentina Vladimirovna Tereshkova became the first
woman to do the same, orbiting Earth forty-eight
times aboard the Vostok 6 spacecraft over the course
of three days. Prior to becoming a space-farer,
Tereshkova had worked as a textile worker, but her
interest in skydiving, coupled with Gagarin's achievement, had led her to volunteer for the Soviet space
programme in 1961. Following eighteen months of
training, she was the only woman of five on the course
to make the grade. On her return to Earth, Tereshkova
was inundated with letters, including this one written
by a twenty-seven-year-old woman with whom she
shared a name. Valentina Vladimirovna Zorkina was a
senior machinist in an oil refinery in the Russian city of
Novokuibyshevsk.*

THE LETTER

Good day, Valentina Vladimirovna!

Dear Valyusha, greetings from your name-sharer, Valentina Vladimirovna Zorkina. I am infinitely happy for you that you had the great honour to make a heroic space flight, the first in the world by a girl from the Soviet Union.

Dear Valyusha! I was staggered by your achievement. I envy you your hardy, persistent nature. I am proud of you. I was so pleased when they began to congratulate you and me, because I am also Valentina Vladimirovna. And I shared your joy. I have no words warm or sincere enough to thank you for your achievement. A huge big thank you. This is the kind of heroic achievement that not everyone could carry out, and not everyone will get such an honour. Only those who have earned it through their painstaking work. I am so proud of you, our Soviet girl.

Dear Valyusha, I have read your biography and it reminded me in part of my own biography. I was also born in the countryside in 1936. My father drove tractors and combine harvesters. We lived in a small house in the country in which we had only one bed, a table, a trunk, a stove and *galanok*. Seven of us lived there. My father went off to war and didn't return. He died there. And we five girls remained with mama. She brought us up on her own. What torment we suffered during the war and in the postwar years. I well remember how we lived then. We ate grass, rotten potatoes, acorns, wild leaves, onions. It was good that we had our own cow so we could have milk. We had nothing to wear on our feet. We had to walk in winter to school from our village—3 kilometres there and 3 kilometres back—in bast [basket] sandals and in one of dad's jackets tied round the waist with string, right up to fifth class. I finished school with merit. I was the last in the family. It took so much strength from my dear mama to raise us all. I owe her an immeasurable debt. And still I finished ten-years schooling and then studied at technical college for another year where I got a diploma. I am now working as a senior machinist at the pumping station in the oil refinery in the town of Novokuibyshevsk. I am raising my daughter Liudochka and I am thinking of going to technical college. I work eight-hour shifts and want to study further. My mama has been living with me for a year now. She's very old, sixty-eight.

Dear Valyusha, I am so proud that you are such a good, sympathetic girl who will help people in their hour of need. Every day I listened to the broadcasts from Moscow and like everyone worried about your safe flight and your safe return to earth. Hooray, you made it back! It's so good that everything went well. Congratulations, Valyusha, on your heroic flight and successful landing. Thank you so very much. Lots of kisses.

Dear Valyusha, I hope you can find a free minute to reply to my letter. How did you feel during the flight? How's your health? Please send me a photograph if you can. I beg you. Say a big hello to Bykovsky, your brother in the heavens.

I wish you a happy life and success in your personal life. And good health. A big thank you to your mama for raising such a daughter. All the women at work say hello. And hello from my husband, Fedya, and my elderly mom too. I sincerely thank you for your feat and am infinitely proud of you. It is such a joy that there are no boundaries, no limits to it. Once again many thanks, my dear Valentina.

I'll stop writing here. Goodbye for now. I look forward to hearing from you.

Kissing you robustly and so proud of you.

With warmest regards,

Valentina Vladimirovna Zorkina

LETTER 18
A HIGHLY CIVILIZED AND INTELLIGENT
RACE OF BEINGS
Alexander Graham Bell to Mabel Hubbard Bell
29 November 1909

Percival Lowell was a businessman and astronomer whose vast wealth enabled him to found the Lowell Observatory in Arizona, which, in 1930, was used to discover Pluto. But Lowell's passion was the study of Mars – more specifically, he was convinced that the red planet was home to a visible network of canals, first spotted by Italian astronomer Giovanni Schiaparelli in 1877. Lowell believed these must have been built by an intelligent civilisation for purposes of irrigation. So convinced was he of this that he published three books about Mars, attracting many followers, including the inventor of the telephone, Alexander Graham Bell. In 1909, having recently read an article of Lowell's on the subject of Venus and its inhospitability, Bell wrote to his wife, Mabel, with some thoughts.

THE LETTER

Mrs. Alexander Graham Bell,
Twin Oaks, Washington, D. C.

My dear Mabel:

I send you a few Sunday thoughts from the House-boat:—

Intelligent Dust

Nov. 28: —The world is about 8000 miles in diameter. It would therefore take 7,040,000 men, each six feet tall, standing one on top of the other, to equal the diameter of the world.

From this it follows that a man is much smaller in proportion to the globe on which he stands, than a cheese-mite is to the cheese he inhabits.

Reduce the world to the size of a cheese. Consider it as a globe one foot in diameter. What then would be the size of a man?

He would be less than 1 seven-millionth of a foot in height about the 1/586675th part of an inch:— An entirely insignificant individual, who would be invisible even under a powerful microscope.

Who would dream of the living dust of a cheese as capable of thought; and yet we ourselves constitute mere dust upon the surface of the world — but <u>intelligent dust</u> at that.

The Neighboring Worlds

<u>Nov. 28</u>: — I am much interested in Percival Lowell's article upon Venus in the Popular Science Monthly for December; and especially in the consequences he deduces from the supposed fact that Venus keeps one face constantly turned towards the Sun, as the Moon keeps one face constantly turned towards us.

In Venus we have a world very like our own. She is about the same size and is surrounded by an atmosphere like ours. She is our next-door neighbor, the nearest to us of all the Planets, and most like the Earth in every respect. The period of her rotation, however, has been a vexed question with Astronomers for a long time past. If, as has often been stated, she rotates once in twenty-four hours, the conditions upon Venus would be so similar to those upon our Earth, as to be quite consistent with the support of similar life there. No signs of life, however, have so far been observed; and now Professor Lowell comes forward with a statement concerning rotation that almost deprives us of hope.

Mars

Mars, our next-door neighbor on the other side, does rotate in about the same period as the earth (24 hours).

He is smaller than the Earth, and has an atmosphere much less dense, so that the conditions are less favorable than here for the existence of life. Still there are indications of life there and of intelligent life too.

We can see the snow-covered areas at the Poles through our telescopes, and watch them melt away when exposed to the Sun. We can see changes of color in some parts of the planet that are suggestive of a process of vegetation. But the real interest of Mars lies in the peculiar markings known as "canals" which suggest the possibility that intelligent beings inhabit Mars who are capable of planning and constructing irrigation works upon a gigantic scale.

Of course the irrigation canals themselves could not be visible from here; and the straight lines observed, whatever they may be, are certainly hot canals. The finest line upon Mars which could be seen from the earth would be several miles in width. All that we could hope to see therefore in connection with an irrigation canal, would be the strip of vegetation bordering upon the canal. A strip of irrigated land, say four or five miles wide, might be seen from the earth as a fine and narrow line; and would resemble one of the so-called "canals" of Mars.

Now what do we actually see upon Mars? We can see the melting of the Polar snow-cap in the Arctic region. Then, as the snow begins to melt, the "canals" begin to appear: First near the Polar area and afterwards extending further down to the South. Professor Lowell translates this to mean that when the ice-cap melts the inhabitants of Mars conduct the water to their desert lands, and irrigate their crops.

The straight lines, or "canals", when first observed, are of a darkish green color, the supposed color of the crop; but as the season advances they change color and become reddish brown, and the crops are supposed to be ripe. Then, following this, the so-called "canals" completely disappear. This may be interpreted to mean that the crops have been harvested, and that Thanksgiving day has arrived. Next year the canals reappear at the same season, and again go through the same series of changes.

The arrangement of these numerous straight lines, or "canals", is such <u>as to indicate design</u>. It is a little difficult to say why; but a glance at a map of Mars will show what I mean.

If we look down upon an American City like Washington D. C. (which was planned as a whole before it was built), we get a map-like effect in which, at a sufficiently great distance away, the parks and houses would appear as blotches of color intersected by a system of lines representing the streets. We could not possibly mistake the pattern formed by these lines for a natural growth; we would at once recognize design in their arrangements relatively to one another. In a similar manner the arrangement of lines upon Mars are suggestive of design.

It is difficult to believe that a line of several hundred miles in length, and perfectly straight from end to end, could be a natural phenomenon. But we find hundreds of such lines upon Mars, extending in every direction, and all connected together apparently in a purposive manner.

Assuming them to be natural formations we would expect them to intersect in places; but, in accidental crossings of this kind we would rarely find more than two lines intersecting at the same point. On Mars, however, it is quite the usual thing for a number of lines to come together at the same point: There is nothing exceptional about the matter at all.

The whole arrangement of lines is so strongly suggestive of an artificial state of things, that we cannot avoid the feeling that we may here be looking down upon the results of the work of intelligent beings inhabiting Mars.

The ability to run a line a thousand miles long, exactly in a straight line, involves mathematical ability of the highest order, and a knowledge of the art of surveying. The running of such a line could probably not be done without reference to outside bodies, like the Sun, the Stars, and the Moons of Mars, thus involving a knowledge of Astronomy.

Great engineering abilities are required, and the highest grade of intellect, to carry out the vast engineering scheme which we seem to see in operation on Mars. This is no less a project than the utilization of the ice of the Arctic regions as a source of water supply for the whole globe, and the distribution of the water to the arid regions for the purposes of irrigation.

If the intersecting straight lines described by Schiaparelli, Lowell, and others, really exist as shown in their drawings, there is no escape from the conviction that Mars is inhabited by a highly civilized and intelligent race of beings carrying on a process of agriculture, and wringing subsistence from the desert by water brought from the Polar regions to irrigate the land.

Venus

It has always seemed strange that we should find indications of life upon the distant planet Mars, where the conditions are very different from those upon Earth; and yet fail to find them on our nearest neighbor, Venus, a sister planet as like the Earth as a twin.

We have hitherto supposed that the reason for our failure may have lain in a difficulty of observation. When Venus is closest to us she is between us and the Sun, and her dark side presents itself to our view. When her bright side is turned this way the Sun is between us, and we cannot see. Mars, on the other hand, presents his bright side at his closest approach; and the net result is that we know a great deal more about Mars than about our closest neighbor, Venus.

Now comes Lowell with the statement that Venus does not spin rapidly upon an axis like the Earth and Mars; but, like Mercury, keeps one side constantly presented to the Sun.

From this he draws interesting, but most disappointing conclusions.

He pictures to us a world scorched upon one side, and frozen on the other; and incapable of any life like that we know. The oceans, he says, have evaporated on the heated side, and have been deposited, as ice and snow, on the other. Cold hurricane winds plough their way into the heated hemisphere from every direction, laden with dust and stones, and tear up the ground into deep radial furrows that can be dimly seen in the telescope of Flagstaff Observatory. The heated air on this hemisphere, he says, overflows above on to the frozen darkened side; but carries no water with it to be deposited there, for the oceans on the heated side have long since been dried up.

There is eternal day there on the heated side, without a cloud to moderate the heat of the Sun: No water, no life, nothing but a desert of scorching sand. On the frozen side we have eternal night, continents of ice and snow, and a cold far below the Arctic temperatures of the Earth, and approaching the absolute zero of space. What a scene of desolation, where we had hoped to find a living and breathing world.

Scientific men imagine that all the planets that move around the Sun were once spinning rapidly upon their axes like tops; but that the tidal effects produced by the attraction of the Sun, and by their Moons, acted as a brake to slow down the rotation. It has long been surmised that Mercury, the planet nearest to the Sun, has already slowed down to such an extent as to make only one rotation in the course of a year, thus keeping one face constantly turned to the Sun; and the next planet, Venus, is now reported to be in the same condition.

The third planet in order from the Sun, is our own inhabited world; and there is evidence that the rotation of the Earth is gradually slowing down. Is our world then to be the next to meet the fate of Mercury and Venus? This is the thought that gives special interest to Lowell's researches.

Of course we need hardly be troubled about the matter at the present time at least on our own account. Nor will our descendants be in any danger for many generations yet to come. It is some comfort to us to know that it will take many thousands of years for the Earth to reach the ultimate condition imagined, in which the Sun will stand still in the heavens, and the succession of days and nights will be no more.

Yet, if Astronomers are right, that time will come; and one-half of the Earth will experience an endless scorching day, and the other half an equally endless Arctic night. Where then will be the human race? Will any survive to perpetuate their kind; or will all life disappear forever from the stricken world? To Venus we look for light, but she only turns her dark side this way.

Though one side of Venus be scorched, and the other frozen, surely there must be a mean somewhere between these opposite conditions. Surely there must be a narrow belt around the planet, between the frozen and heated sides constituting a temperate zone where life may possibly exist: A region, where the Sun would be forever upon the horizon, without either setting to freeze, or rising to roast, the living things that might take refuge there.

Upon the existence of this narrow ring of life I pin my faith; and I do not yet despair of the safety of the human race.

AGB

'WE OURSELVES
CONSTITUTE MERE
DUST UPON THE
SURFACE OF THE WORLD
– BUT INTELLIGENT
DUST AT THAT.'

– Alexander Graham Bell

I'LL BE WATCHING OVER YOU

Jerry Linenger to John Linenger

23 January 1997

Jerry Linenger, a forty-one-year-old American astronaut,
kissed his pregnant wife and fourteen-month-old son
goodbye on 12 January 1997, boarded Space Shuttle
Atlantis and headed for Space Station Mir. There he
was to join two Russian cosmonauts, Vasili Tsibliyev
and Aleksandr Lazutkin. Linenger remained in space for
a record-breaking 132 days, during which time he and
his crew faced a number of problems, the most
dangerous being a raging fire that burned for fourteen
minutes, nearly destroying the space station and all on
board. Linenger wrote many letters home during his
stay, of which these are just three. He returned safely
to Earth on 24 May. His wife gave birth to their
second son a few months later.

THE LETTERS

23 January 1997

Dear John,

I decided before this flight that I was going to be a good father and write to you every day. This is my first attempt at that.

I realize that you are only one year old, and although I exaggerate your talents like any proud father would, I don't think that you can quite read this yet. No problem. When you can, you'll feel good knowing that your father loves you.

Space flight is a dangerous business. I used to be pretty cavalier about it. But just before this launch, I started questioning what I was about to do. You see, I have so, so much to lose now. You and your mother.

I always liked adventures. I remember exhausting the elementary school library of mystery books by someone I think was named Orton. Trying to figure out the ending before the ending. Anticipating. Observing the situation, and trying to predict the outcome. Reading about people who were in unusual situations, and studying how they were challenged, and how they responded.

Anyway, that curiosity characteristic is what got me on this space station. Oh sure, I went to lots of schools, did pretty well in our great United States Navy, and went through all the mechanics of the application and interview process. But the basic trait of insatiable curiosity is what drove me through all of that.

Space is a frontier. And I'm out here exploring. For five months! What a privilege!

But, I sure do miss you. Want most of all to see you come stumbling around the corner, bellow out your big laugh when I give my "surprised to see you" look, and watch you stumble back out of the room to do the same to mom in the other room. You are the best son in the world, John.

You know, although I am up here floating above earth, I am still an earthling. I feel the pain of separation, the pride of a father, and the loneliness of a husband away from his wife like an earthling. And maybe even a bit more acutely.

Good night, my son. I'll be watching over you.
Dad

* * *

Hello John,

People often ask me what I miss.

You and Mommy, of course. Likewise, family and friends.

But I also miss fresh air blowing in my face. Green, green grass and swaying trees. Birds chirping. Tulips popping up in spring.

Taking hot showers. Lying on the couch. Falling asleep with two big pillows surrounding my head. Diving into the swimming pool after a long, hot run.

Tinkering in the garden. Looking out over the lake as the sun sets. Feeling the warmth of the sun. Gliding across the water in a kayak with fish jumping in my wake.

Pretzels. The smell of popcorn, or better yet, homemade bread baking. Dinner conversations with Mommy. Cuddling.

Silent nights. Crickets. Waves pounding on the shore. Walking barefoot in the sand. Walking. Holding hands.

Basically, I miss the elemental things of Earth that we are blessed with each day on the planet but often take for granted.

After I land, my eyes will be opened as wide as

yours always are, John, as I rediscover the little pleasures. Father and son, holding hands and out adventuring together.

Rest up. We will be busy together. Good night.

Love,

Dad

* * *

23 May 1997

Dear John,

I have changed post offices. This letter is being sent down from the space shuttle *Atlantis*, and in a day or so I will be home.

We closed the hatch last night between the Mir space station and the shuttle in order to be prepared for our early-morning departure. Following a gentle push-off, we began intermittently firing our thrusters. The bursts made loud bang-bang-bang sounds, similar in abruptness to cannon firing. As we moved away, the Mir became smaller, then smaller still. Finally, it was so diminished in size that the space station appeared to be nothing more than a rather insignificant blinking light among the stars.

Surprisingly, I felt very little emotion when leaving my home of the last one hundred and twenty-two days behind. I suppose that I just felt like my time was up, that I had done my duty, and that it was time for me to go. This is in stark contrast to the very strong emotion that I felt upon first seeing the space shuttle *Atlantis* arrive a few days ago. When I saw *Atlantis* approaching the Mir, I felt pure elation, pure unbridled joy.

I have an image in my mind of the time you took your first steps. I would move two steps away from you, leaving you standing and holding on, precariously, to the edge of the sofa. You would look at me with questioning eyes. Your eyes reflected what I am sure that you were asking yourself: "Can I do this, or will I fall?" After I would encourage you with a reassuring word or gesture, you would muster up your courage, let go, and walk to me.

Reporters keep asking me whether, after landing, I plan to get out of *Atlantis* on my own power or be carried off. I will be trying my best to follow in your footsteps, John. I will be giving it everything that I have to walk, or crawl, or do whatever it takes, but to do it on my own, just like you did.

I hope to be standing and holding you in a day or two, John.

Love,

Dad

CONFINED TO EARTH, WE HAVE REACHED OUR LIMITS

Isaac Asimov to Adlai Stevenson

18 November 1979

Author and biochemist Isaac Asimov was just three years old when his family moved from Russia to the United States in 1923. He would remain there for the rest of his life. In a prolific career spanning forty years, Asimov wrote more than five hundred books, many of which – like his epic, genre-defining Foundation series of novels – could be categorised as science fiction. But he also wrote a number of educational non-fiction titles such as Isaac Asimov's Guide to Earth and Space, in which he explained concepts sometimes found in his novels. Such was his expertise in all things astronomical that Asimov would sometimes be asked by politicians to offer his thoughts on related matters. In 1979, he wrote this letter to Democratic senator Adlai Stevenson in response to such a request.

THE LETTER

New York, N.Y.
November 18, 1979

Hon. Adlai E. Stevenson,
U.S. Senate,
Washington, D.C.

Dear Senator Stevenson: I have, in many articles, made it clear that I believe the exploration and exploitation of space would not only be of advantage to the United States, but is absolutely imperative if human civilization is to be preserved.

It is from space that we will get the material resources, the energy, and the gains in technology that will be needed to allow humanity to continue to grow and expand. Confined to Earth, we have reached our limits.

Furthermore, international cooperation must be made a reality, and war must be abolished and this can only be done if we lose our localisms in a great, unifying project. Space offers the only such project.

I suspect that you share these beliefs with me. You, however, intimately involved with the nuts-and-bolts of politics and economics are much better suited than I am to chart the exact course to these goals.

I am an ivory-tower writer (dreamer, perhaps) and I feel completely unqualified to advise you in these practical matters.

Isaac Asimov

'[SPACE EXPLORATION]
IS ABSOLUTLEY
IMPERATIVE IF HUMAN
CIVILIZATION IS TO BE
PRESERVED.'

— Isaac Asimov

LETTER 21
THE PROVERBIAL 'REALLY GOOD'
SCIENCE-FICTION MOVIE
Stanley Kubrick to Arthur C. Clarke
31 March 1964

*Esteemed filmmaker Stanley Kubrick initiated contact
with author Arthur C. Clarke by way of this letter, in
which Kubrick declared an interest in the two collabo-
rating. Clarke was immediately keen – so much so that
just three weeks later, on 22 April, the pair met at the
Plaza Hotel in New York and, according to Clarke,
'talked for eight solid hours about science fiction'. Four
years later, the ground-breaking result of their partner-
ship, 2001: A Space Odyssey, was released, a truly epic
movie about space travel, human evolution and extra-
terrestrial life, spanning time and space. It is
considered a landmark in cinema.*

THE LETTER

SOLARIS PRODUCTIONS, INC

March 31, 1964

Dear Mr Clarke:

It's a very interesting coincidence that our mutual friend Caras mentioned you in a conversation we were having about a Questar telescope. I had been a great admirer of your books for quite a time and had always wanted to discuss with you the possibility of doing the proverbial "really good" science-fiction movie.

My main interest lies along these broad areas, naturally assuming great plot and character:

1. The reasons for believing in the existence of intelligent extra-terrestrial life.
2. The impact (and perhaps even lack of impact in some quarters) such discovery would have on Earth in the near future.
3. A space-probe with a landing and exploration of the Moon and Mars.

Roger tells me you are planning to come to New York this summer. Do you have an inflexible schedule? If not, would you consider coming sooner with a view to a meeting, the purpose of which would be to determine whether an idea might exist or arise which would sufficiently interest both of us enough to want to collaborate on a screenplay.

[. . .]

Incidentally, "Sky & Telescope" advertise a number of scopes. If one has the room for a medium size scope on a pedestal, say the size of a camera tripod, is there any particular model in a class by itself, as the Questar is for small portable scopes?

Best regards,
[Signed]
Stanley Kubrick

LETTER 22
DEAR SON
Marion Carpenter to Malcolm Scott Carpenter
May 1962

Malcolm Scott Carpenter was selected by NASA as one of seven astronauts to be involved with Project Mercury, the United States' programme focused entirely on launching humans into space. At Cape Canaveral in Florida on 24 May 1962, Carpenter boarded the Aurora 7 spacecraft and circled Earth three times. He was just the second American to orbit his home planet. On the eve of his historic journey, Carpenter's father, Marion, proudly wrote him this letter.

THE LETTER

M. Scott Carpenter
Palmer Lake
Colorado

Dear Son,

Just a few words on the eve of your great adventure for which you have trained yourself and anticipated for so long—to let you know that we all share it with you, vicariously.

As I think I remarked to you at the outset of the space program, you are privileged to share in a pioneering project on a grand scale—in fact, the grandest scale yet known to man. And I venture to predict that after all the huzzahs have been uttered and the public acclaim is but a memory, you will derive the greatest satisfaction from the serene knowledge that you have discovered new truths. You can say to yourself: this I saw, this I experienced, this I know to be the truth. This experience is a precious thing; it is known to all researchers, in whatever field of endeavour, who have ventured into the unknown and have discovered new truths.

You are probably aware that I am not a particularly religious person, at least in the sense of embracing any of the numerous formal doctrines.

Yet I cannot conceive of a man endowed with intellect, perceiving the ordered universe about him, the glory of the mountain top, the plumage of a tropical bird, the intricate complexity of a protein molecule, the utter and unchanging perfection of a salt crystal, who can deny the existence of some higher power. Whether he chooses to call it God or Mohammed or the Turquoise Woman or the Law of Probability matters little. I find myself in my writings frequently calling upon Mother Nature to explain things and citing Her as responsible for the order of the universe. She is a very satisfactory divinity for me. And so I shall call upon Her to watch over you and guard you and, if she so desires, share with you some of her secrets which she is usually so ready to share with those who have high purpose.

    With all my love,
        Dad

LETTER 23
MISS MITCHELL'S COMET
William Mitchell to William Cranch Bond

3 October 1847

*Born in 1818, Maria Mitchell was the first professional
female astronomer in the United States, the first
woman elected to the American Academy of Arts and
Sciences, and the first woman to become a professor
of astronomy – a post she held at Vassar College in
New York from 1865 until 1888. She also, on 1 October
1847, discovered C/1847 T1, a comet also spotted by
Italian astronomer Francesco de Vico two days later,
and which was temporarily named after him until
news of Mitchell's sighting reached Europe. Shortly
after the event, her father, acutely aware of his
daughter's achievement and the need to report it
immediately, alerted the head of the Harvard College
observatory by letter. C/1847 T1 is now known as
'Miss Mitchell's Comet'.*

THE LETTER

Nantucket, 10 mo. 3d, 1847.

My dear Friend,—
I write now merely to say that Maria discovered a
telescopic comet at half past ten on the evening of
the first instant, at that hour nearly vertical above
Polaris five degrees. Last evening it had advanced
westwardly; this evening still further, and nearing
the pole. It does not bear illumination, but Maria
has obtained its right ascension and declination,
and will not suffer me to announce it. Pray tell me
whether it is one of George's; if not, whether it has
been seen by any body. Maria supposes it may be
an old story. If quite convenient, just drop a line to
her; it will oblige me much. I expect to leave home
in a day or two, and shall be in Boston next week,
and I would like to have her hear from you before
I can meet you. I hope it will not give thee much
trouble amidst thy close engagements.

Our regards are to all of you, most truly,
William Mitchell

LETTER 24
OTHERS BELIEVE A POET OUGHT TO GO
TO THE MOON

Julian Scheer to George M. Low
12 March 1969

Approximately 650 million people around the world watched with mouths agape as Neil Armstrong descended the lunar module ladder and stepped onto the surface of the Moon on 20 July 1969. This was a moment of unparalleled importance, entirely without precedent. Never before had a single person commanded the attention of so many. It was at that moment, as he took humankind's first steps on the Moon, that Armstrong uttered a string of words that would soon take their place in history books: 'That's one small step for man. One giant leap for mankind.' Then followed by Buzz Aldrin's utterance: 'Beautiful view. Magnificent desolation.'

Armstrong later said of his choice of words, 'I thought about it after landing,' and both astronauts have since made clear that their words were their own, uncoached by NASA. This fascinating letter, written months before launch by NASA spokesman Julian Scheer, discussed exactly that.

THE LETTER

NATIONAL AERONAUTICS AND SPACE
ADMINISTRATION
WASHINGTON. D.C 20546

March 12, 1969

Mr. George M. Low
Manager
Apollo Spacecraft Program
NASA Manned Spacecraft Center
Houston, Texas 77058

Dear George:

It has come to my attention that you have asked someone outside of NASA to advise you on what the manned lunar landing astronauts might say when they touch down on the Moon's surface. This disturbs me for several reasons.

The Agency has solicited from within NASA any suggestions on what materials and artifacts might be carried to the surface of the Moon on that historic first flight. But we have not solicited comment or suggestions on what the astronauts might say. Not only do I personally feel that we ought not to coach the astronauts, but I feel it would be damaging for the word to get out that we were soliciting comment. The ultimate decision on what the astronauts will carry is vested in a committee set up by the Administrator; the committee will not, nor will the Agency by any other means, suggest remarks by the astronauts.

Frank Borman solicited a suggestion from me on what would be appropriate for Christmas Eve. I felt – and my feeling still stands – that his reading from the Bible would be diminished in the eyes of the public if it were thought that NASA pre-planned such a thing. I declined both officially and personally to suggest words to him despite the fact that I had some ideas. I believed then and I believe the same is true of the Apollo 11 crew that the truest emotion at the historic moment is what the explorer feels within himself, not for the astronauts to be coached before they leave or to carry a prepared text in their hip pocket.

The Lunar Artifacts Committee, chaired by Willis Shapley, asked that all elements of NASA consider what might be carried on Apollo 11. I know that General Phillips has properly reiterated the request by asking all elements of Manned Flight to suggest things, but it was not the desire or intent of the committee to broaden the scope of the solicitation to verbal reactions.

There may be some who are concerned that some dramatic utterance may not be emitted by the first astronaut who touches the lunar surface. I don't share that concern. Others believe a poet ought to go to the Moon. Columbus wasn't a poet and he didn't have a prepared text, but his words were pretty dramatic to me. When he saw the Canary Islands he wrote, "I landed, and saw people running around naked, some very green trees, much water, and many fruits."

Two hundred years before Apollo 8, Captain James Cook recorded while watching the transit of Venus over the sun's disk, "We very distinctly saw an atmosphere or dusky shade around the body of the planet."

Meriwether Lewis, traveling with William Clark, recorded, "Great joy in camp. We are in view of the ocean, this great Pacific Ocean which we have been so long anxious to see, and the roreing [sic] or noise made by the waves brakeing [sic] on the rockey [sic] shore may be heard distinctly."

Peary was simply too tired to say anything in 1909 when he reached the North Pole. He went to sleep. The next day he recorded in a diary. "The pole at last. The prize of three centures [sic]. I cannot bring myself to realize it. It seems all so simply and commonplace."

The words of these great explorers tell us something of the men who explore and it is my hope that Neil Armstrong or Buzz Aldrin will tell us what they see and think and nothing that we feel they should say.

I have often been asked if NASA indeed plans to suggest comments to the astronauts. My answer on behalf of NASA is "no".

I'd appreciate your comments.

Regards,

Julian Scheer

Assistant Administrator

for Public Affairs

Cc: Neil Armstrong

Mike Collins

Buzz Aldrin

'THE WORDS OF THESE
GREAT EXPLORERS TELL
US SOMETHING OF THE
MEN WHO EXPLORE.'

— Julian Scheer

LETTER 25
IT'S THE TRIP OF A LIFETIME!

Buzz Aldrin to Barry Goldman

25 September 1997

*Twelve years after graduating from the US Military
Academy in West Point and having flown dozens of
combat missions in the Korean War, Edwin E. 'Buzz'
Aldrin was selected by NASA to become an astronaut
in 1963. Three years later, he orbited Earth aboard the
Gemini 7 spacecraft. Famously, after landing the lunar
module on the Moon's surface in 1969, Aldrin became
the second human to set foot on it, directly after Neil
Armstrong. Aldrin received many thousands of letters
during his illustrious career, from all manner of people.
In 1997, he responded to a professor at the University
of Maryland with this letter.*

THE LETTER

Buzz Aldrin

September 25, 1997
Barry Goldman, Faculty
University of Maryland

Dear Mr. Goldman,

I am writing to you to share some of my personal ideas and thoughts about my experiences related to the moon landing.

I have often described the moon as a "magnificent desolation." Its rocky horizon curved against the deep black of space, making it perfectly obvious that we were standing on a ball spinning through the universe.

When I planted the American flag on the dusty surface of the moon, I had an unusual thought: A billion people were watching me on television. Human beings had never been farther away than we were nor had more people thinking about them!

I think the spirit and the sense of involvement exhibited by the numbers of people who remember where they were when that event happened make it even more apparent to me over the years that the moonwalk added value to the lives of all the people who participated in it. Every person felt good about the nation achieving it – that the world, that humanity could have done this.

I have snapshots of myself on the moon that will always remind me of that strange and fascinating place. Someday in the future as people are mulling over their vacation plans, I hope they'll choose to fly into space. It's the trip of a lifetime!

Regarding your questions of space exploration in 50 years: all of the rationales reduce to one simple truth – we will walk on Mars in the spirit and wonder that sets our species apart.

Sincerely,

Buzz Aldrin

Ad Astra via Mars!

APOLLO XI

'SOMEDAY IN THE
FUTURE AS PEOPLE ARE
MULLING OVER THEIR
VACATION PLANS, I
HOPE THEY'LL CHOOSE
TO FLY INTO SPACE.'

– Buzz Aldrin

LETTER 26
I MAY BE OF SOME USE TO THE PRESIDENT

Ray Bradbury to Arthur Schlesinger Jr; John F. Kennedy to Ray Bradbury

April/June 1962

The Space Race was a thrilling time for space enthusiasts, filled as it was with previously unimaginable achievements and firsts, beginning in 1957 with the Soviets' launch of Sputnik 1. From the creation of NASA in 1958 through to the various satellites, humans and animals that were thrust into the sky by two hyper-competitive teams of unspeakably intelligent people, each and every move by the Soviet Union and the USA heightened the tension. It is no wonder, then, that a science-fiction author of Ray Bradbury's calibre and experience would want, somehow, to get involved. In 1962, weeks before industry experts joined Kennedy at the Conference on the Peaceful Uses of Space in Seattle, Bradbury plainly offered his services to the Administration by way of this letter to Arthur Schlesinger, advisor to the President, along with copies of his books. A brief reply of thanks arrived from Kennedy two months later. Bradbury's offer was not mentioned.

THE LETTERS

April 30th, 1962

Mr. Arthur Schlesinger, Jr.
The White House
Washington, D.C.

Dear Arthur: (If I may, please)
You are probably wondering what happened to me
after my phone calls last Thursday. Friday was one
of my lost days, traveling all over the 400 square
mile Los Angeles territory, lecturing. So, rather than
try to telephone again, I decided on this letter.

Under separate cover, airmail, I'm sending on
two complete sets of all my books. One set is autographed to you. The other is autographed to the
President and Mrs. Kennedy.

I'm sending these for various reasons:

We are moving deeper into the Space Age, a
tremendously exciting time for me, as you can
imagine, since I started writing about it when I was
twelve. Now, I hear that there will be a Conference
on the Peaceful Uses of Outer Space, attended by
President Kennedy and 1500 of the top scientists in
the field of space travel from all over the world, in
Seattle. A few days later, I'm scheduled to appear as

a special Guest at the Seattle Fair. Our lives move closer together with such events.

Somewhere along the line I may be of some use to the President, or to you, or to others working with the President. It may be in the field of writing. Or it might be in some other act I could perform for any one of you.

I'm now finishing a 14 minute film, ICARUS MONTGOLFIER WRIGHT, based on man's age-old desire to fly. This short has been made with Format Films here, in semi-animation color, and when it is finished I hope to show it to you, and, perhaps, the President.

Starting in a few weeks, I will begin to give a series of short talks on NBC-MONITOR-RADIO, coast to coast, dealing with every aspect of the Space Age: music, poetry, painting, architecture, psychology, etc.

My one-act space-age plays will be staged at the Royal Court theatre in London this summer.

Jean-Louis Barrault, the French actor-mime-director, will open my THE MARTIAN CHRONICLES on-stage in Paris this winter, at the Odeon.

Francois Truffaut, the director of THE 400 BLOWS, will film my FAHRENHEIT 451 this autumn, in France.

So you see I have thought, written, and now am doing much, in many directions, about the Space Age.

I have marked those stories, particularly, in THE GOLDEN APPLES OF THE SUN and THE DAY IT RAINED FOREVER, which I think might fascinate you and the President.

There are many people who know the facts about the various projects headed toward space. There are only a few who interpret, aesthetically or otherwise, our entire purpose in Space. I have tried to do this quite often.

If the President thinks of any way I might serve him or the Government, with my particular talent, I would be glad to help promote the Space Age as we would all like to see it promoted, as a motion toward peace and survival.

If you wish, you may show this letter to the President.

My cordial best to you,
Ray Bradbury

* * *

June 21, 1962

Dear Mr. Bradbury:
I was delighted to receive through Arthur Schlesinger your four volumes. I am happy to have available so generous a selection of your writings

which will allow me to indulge simultaneously both fantasy and scientific reality. I have heard from many sources of your talent, and welcome this chance to experience it at first hand.

With all good wishes,

Sincerely,

John F. Kennedy

'I WOULD BE GLAD TO
HELP PROMOTE THE
SPACE AGE AS WE
WOULD ALL LIKE TO SEE
IT PROMOTED.'

— Ray Bradbury

LETTER 27
MAKE PLUTO A PLANET AGAIN
Cara Lucy O'Connor and NASA
April 2017

For seventy-six years, Pluto was considered the ninth planet in our solar system. Discovered by American astronomer Clyde Tombaugh on 18 February 1930, this entity made up of ice and rock can be found in the Kuiper belt, a ring-shaped mass of objects circling the Sun, of which Pluto is the largest, despite its measuring just half the width of the United States. A single day on Pluto lasts 153 hours. In 2006, Pluto was demoted – at least in the eyes of the public – and reclassified as a dwarf planet by the International Astronomical Union, a controversial decision causing letters of complaint to reach NASA to this day. In 2017, this missive arrived at the organisation from a young girl named Cara. For her troubles, she received not one but two replies.

THE LETTERS

To NASA

My name is Cara Lucy O'Connor. I am from Cork in Ireland. I am 5 years old and I am in Junior Infants in Glasheen Girls School. Recently I have become very interested in astronomy because I started watching videos about Outer Space.

I am writing to you (with the help of my teacher Miss O'Donovan because I'm still learning to spell all the words) because I have a concern. I listened to a song and at the end of it the song said "Bring Pluto Back" and I would really like that to happen. Pluto was reclassified in 2006. Pluto used to be a planet and I think that was fair but it isn't fair that Pluto isn't a planet anymore. Now it is a Dwarf Planet. A dwarf planet is a type of planet that isn't big enough to clear its orbit. Some of the dwarf planets are located in the Kuiper belt and there are rocks and asteroids there as well.

I really think Pluto should be a main planet again like Mercury, Venus, Earth, Mars, Jupiter, Saturn, Uranus & Neptune, because in one video I watched called "Let's go meet the planets", Pluto was included at the very end. I listened to a few songs about Pluto as well, and in one of the videos people were dressed as different planets, and Pluto was put in the trash can and was scared by planet Earth. This was really mean because no one or no planet or dwarf planets should be put in the trash can.

I would really like it if everyone at NASA could please change your minds and make Pluto a planet again.

When I turn 6 I am hoping to discover my own planet and call it Planet Unicorn so I would like to visit all of the main planets including Pluto. I am hoping that one day I can become an astronaut and work for ye at NASA but you need to fix this problem for me.

I would really love if you could write back to me soon

Lots of love

Cara Lucy O'Connor

PS I am Goldillocks [sic] in our Summer show if you would like to come see it.

* * *

SOUTHWEST RESEARCH INSTITUTE

Dear Cara,

I got your letter, and while I don't work for NASA

I do work on NASA's New Horizons spacecraft that flew past Pluto in 2015. Thank you for your letter, and your interest in Pluto. I think it's very good when young people, especially girls, are interested in the planets. I grew up near London, and I was interested in the planets too when I was young.

You're right. Pluto was made a Dwarf Planet. Some people I work with think this is right, and some do not. I'm not really sure. On the one hand I think it is good when things are grouped correctly, it helps us understand them all better. For example, if we understand something better about a fish then that helps us learn about all fish, not just that one. However, I think Pluto is important to a lot of people, and they think that because we used to think of it as a planet it should remain so. I think Pluto is very important, but Pluto probably doesn't care what silly people on the Earth call it. So I just call it my favorite world and let other people argue about it.

I think Planet Unicorn would be amazing. I think it would have a lot of rainbows, and anyone that lived there would be happy. Lots of planets are being discovered around other stars at the moment, so I think if you work hard at school (particularly in Maths and Science) you could discover Planet Unicorn soon.

Keep up the good work, we need people to
imagine what discoveries are still out there.

Yours truly,

Dr. Carly Howett

Deputy Principle Investigator of the Ralph instru-
ment on NASA's New Horizons spacecraft
Acting Assistant Director of the Department for
Space Studies
Senior Research Scientist and Outer Solar System
Section Manager

Southwest Research Institute
Boulder, Colorado, 80302
USA

* * *

NASA

SMD/Planetary Science Division
AUG 01 2017

Dear Cara Lucy.

Thank you for writing and for your interest in
NASA and space! I'm the director of NASA's
Planetary Science Division, which means I

oversee all our missions to the planets, including Pluto.

You are right that Pluto was reclassified as a dwarf planet about 11 years ago. That decision was made by a group of astronomers known as the International Astronomical Union.

I agree with you that Pluto is really cool—in fact, who would have believed that Pluto has a heart? Since our New Horizons spacecraft flew by Pluto, we've learned that Pluto isn't a boring cratered rock. It's a fascinating world that appears to be constantly changing. To me, it's not so much about whether Pluto is a dwarf planet or not; it's that Pluto is a fascinating place that we need to continue to study.

I hope that you will discover a new planet, and I trust that if you continue to do well in school we will see you at NASA one of these days.

Sincerely,

[Signed]

Dr. James L. Green

Director, Planetary Science Division

LETTER 28
IN EVENT OF MOON DISASTER
William Safire to H. R. Haldeman
18 July 1969

It's difficult to imagine a missive more chilling than this one, expertly written by presidential speechwriter William Safire for White House Chief of Staff H. R. Haldeman as the world waited anxiously for Apollo 11 to land safely on the surface of the Moon. It was a contingency plan of sorts, containing a speech to be read out by Nixon to the public, should astronauts Neil Armstrong and Buzz Aldrin fail to return, and instructions for the President to call and inform the 'widows-to-be' of the tragedy. As we now know, the memo was thankfully never needed; all that remains is an eerie reminder that things could have gone terribly wrong, and that those at the top were very much prepared for such an unthinkably bleak eventuality.

THE LETTER

To: H. R. Haldeman

From: Bill Safire
July 18, 1969.

IN EVENT OF MOON DISASTER:

Fate has ordained that the men who went to the moon to explore in peace will stay on the moon to rest in peace.

These brave men, Neil Armstrong and Edwin Aldrin, know that there is no hope for their recovery. But they also know that there is hope for mankind in their sacrifice.

These two men are laying down their lives in mankind's most noble goal: the search for truth and understanding.

They will be mourned by their families and friends; they will be mourned by the nation; they will be mourned by the people of the world; they will be mourned by a Mother Earth that dared send two of her sons into the unknown.

In their exploration, they stirred the people of the world to feel as one; in their sacrifice, they bind more tightly the brotherhood of man.

In ancient days, men looked at stars and saw their heroes in the constellations. In modern times, we do much the same, but our heroes are epic men of flesh and blood.

Others will follow, and surely find their way home. Man's search will not be denied. But these men were the first, and they will remain the foremost in our hearts.

For every human being who looks up at the moon in the nights to come will know that there is some corner of another world that is forever mankind.

PRIOR TO THE PRESIDENT'S STATEMENT:

The President should telephone each of the widows-to-be.

AFTER THE PRESIDENT'S STATEMENT, AT THE POINT WHEN NASA ENDS COMMUNICATIONS WITH THE MEN:

A clergyman should adopt the same procedure as a burial at sea, commending their souls to "the deepest of the deep," concluding with the Lord's Prayer.

'THEY WILL BE
MOURNED BY A
MOTHER EARTH THAT
DARED SEND TWO OF
HER SONS INTO THE
UNKNOWN.'
— Bill Safire

LETTER 29
I AM CLEARLY SUSPECT AND NOT
BELIEVABLE
Neil Armstrong to James Whitman
10 November 2005

In October of 2005, a social studies teacher at an Ohio high school wrote to Neil Armstrong with some questions. This was thirty-six years after Armstrong had become the first human to step onto the Moon, his incredible achievement filmed and beamed, live, into the homes of hundreds of millions of people around the world. Alas, despite the broadcast, a surprising number of conspiracy theorists – Mr Whitman included – have for many years debated the Moon landing, positing that the footage was in fact filmed in a studio much closer to home, and that the crew of Apollo 11 failed to get close to the lunar surface. A number of these sceptics wrote to the Apollo astronauts, but rarely did they receive a response. Mr Whitman, however, was lucky.

THE LETTER

Mr. Whitman,

Your letter expressing doubts based on the skeptics and conspiracy theorists mystifies me.

They would have you believe that the United States Government perpetrated a gigantic fraud on its citizenry. That the 400,000 Americans who worked on an <u>unclassified</u> program are all complicit in the deception, and none broke ranks and admitted their deceit.

If you believe that is possible, why would you contact me, clearly one of those 400,000 liars?

I trust that you, as a teacher, are an educated person. You will know how to contact knowledgeable people who could not have been a party to the scam.

The skeptics claim that the Apollo flights did not go to the moon. You could contact the experts from other countries who tracked the flights on radar (Jodrell Bank in England or even the Russian Academicians).

You should contact the Astronomers at Lick Observatory who bounced their laser beam off the Lunar Radar Ranging Reflector minutes after I installed it. Or, if you don't find them persuasive, you could contact the astronomers at the Pic du

Midi observatory in France. They can tell you about all the other astronomers in other countries who are still making measurements from those same mirrors—and you can contact them.

Or you could get on the net and find the researchers in university laboratories around the world who are studying the lunar samples returned on Apollo, some of which have never been found on earth.

But you shouldn't be asking me, because I am clearly suspect and not believable.

Neil Armstrong

'YOUR LETTER
EXPRESSING DOUBTS
BASED ON THE SKEPTICS
AND CONSPIRACY
THEORISTS MYSTIFIES
ME.'

— *Neil Armstrong*

LETTER 30
HAPPY BIRTHDAY, BUDDY
Neil deGrasse Tyson to NASA
July 2008

On 29 July 1958, prompted by the Soviets' launch of the world's first artificial satellite, Sputnik 1, US President Dwight D. Eisenhower signed the National Aeronautics and Space Act. This piece of legislation created the National Aeronautics and Space Administration, known by most as NASA. Progress was rapid, and two years later the Apollo program was born. Within a decade, the agency had placed men on the Moon. In 2008, when NASA turned fifty, this celebratory letter was written by Dr Neil deGrasse Tyson, noted astrophysicist, author and Frederick P. Rose Director of the American Museum of Natural History's Hayden Planetarium in New York City.

THE LETTER

Dear NASA,

Happy birthday! Perhaps you didn't know, but we're the same age. In the first week of October 1958, you were born of the National Aeronautics and Space Act as a civilian space agency, while I was born of my mother in the East Bronx. So the yearlong celebration of our golden anniversaries, which begins the day after we both turn 49, provides me a unique occasion to reflect on our past, present and future.

I was 3 years old when John Glenn first orbited Earth. I was 7 when you lost astronauts Grissom, Chaffee and White in that tragic fire of their Apollo 1 capsule on the launch pad. I was 10 when you landed Armstrong and Aldrin on the moon. And I was 14 when you stopped going to the moon altogether. Over that time I was excited for you and for America. But the vicarious thrill of the journey, so prevalent in the hearts and minds of others, was absent from my emotions. I was obviously too young to be an astronaut. But I also knew that my skin color was much too dark for you to picture me as part of this epic adventure. Not only that, even though you are a civilian agency, your most celebrated astronauts were military pilots, at a time when war was becoming less and less popular.

During the 1960s, the Civil Rights movement was more real to me than it surely was to you. In fact, it took a directive from Vice President Johnson in 1963 to force you to hire black engineers at your prestigious Marshall Space Flight Center in Huntsville, Ala. I found the correspondence in your archives. Do you remember? James Webb, then head of NASA, wrote to German rocket pioneer Wernher von Braun, who headed the center and who was the chief engineer of the entire manned space program. The letter boldly and bluntly directs von Braun to address the "lack of equal employment opportunity for Negroes" in the region, and to collaborate with the area colleges Alabama A&M and Tuskegee Institute to identify, train and recruit qualified Negro engineers into the NASA Huntsville family.

In 1964, you and I had not yet turned 6 when I saw picketers outside the newly built apartment complex of our choice, in the Riverdale section of the Bronx. They were protesting to prevent Negro families, mine included, from moving there. I'm glad their efforts failed. These buildings were called, perhaps prophetically, the Skyview Apartments, on whose roof, 22 stories over the Bronx, I would later train my telescope on the universe.

My father was active in the Civil Rights movement, working under New York City's Mayor Lindsay to create job opportunities for youth in the ghetto – as the "inner city" was called back then. Year after year, the forces operating against this effort were huge: poor schools, bad teachers, meager resources, abject racism and assassinated leaders. So while you were celebrating your monthly advances in space exploration from Mercury to Gemini to Apollo, I was watching America do all it could to marginalize who I was and what I wanted to become in life.

I looked to you for guidance, for a vision statement that I could adopt that would fuel my ambitions. But you weren't there for me. Of course, I shouldn't blame you for society's woes. Your conduct was a symptom of America's habits, not a cause. I knew this. But you should nonetheless know that among my colleagues, I am the only one in my generation who became an astrophysicist in spite of your achievements in space rather than because of them. For my inspiration, I instead turned to libraries, remaindered books on the cosmos from bookstores, my rooftop telescope and the Hayden Planetarium. After some fits and starts through my years in school, where becoming an astrophysicist seemed at times to be the path of most resistance through an unwelcoming society, I became a professional scientist. I became an astrophysicist.

Over the decades that followed you've come a long way, including, most recently, a presidentially initiated, congressionally endorsed vision statement that finally gets us back out of low-Earth orbit. Whoever does not yet recognize the value of this adventure to our nation's future soon will, as the rest of the developed and developing world passes us by in every measure of technological and economic strength. Not only that, today you look much more like America – from your senior-level managers to your most decorated astronauts. Congratulations. You now belong to the entire citizenry. Examples of this abound, but I especially remember when the public took ownership of the Hubble Space Telescope, your most beloved unmanned mission. They all spoke loudly back in 2004, ultimately reversing the threat that the telescope might not be serviced a fourth time, extending its life for another decade. Hubble's transcendent images of the cosmos had spoken to us all, as did the personal profiles of the space shuttle astronauts who deployed and serviced the telescope and the scientists who benefited from its data stream.

Not only that, I've even joined the ranks of your most trusted, as I served dutifully on your prestigious Advisory Council. I came to recognize that when you're at your best, nothing in this world can inspire the dreams of a nation the way you can – dreams fueled by a pipeline of ambitious students eager to become scientists, engineers and technologists in the service of the greatest quest there ever was. You have come to represent a fundamental part of America's identity, not only to itself but to the world.

So as we both turn 49, and begin our 50th trip around the sun, I want you to know that I feel your pains and share your joys. And I look forward to seeing you back on the moon. But don't stop there. Mars beckons, as do destinations beyond.

Happy birthday, buddy. Even if I have not always been, I am now your humble servant,

Neil deGrasse Tyson

Astrophysicist, American Museum of Natural History

PERMISSION CREDITS

Every effort has been made to trace copyright holders and obtain their permission for the use of copyright material. The publisher apologises for any errors or omissions and would be grateful if notified of any corrections that should be incorporated in future reprints or editions of this book.

LETTER 3 from Carl Sagan to Alan Lomax (1977) from https://www.loc.gov/item/cosmos000113/, Used with permission.

LETTER 4 from Ann Druyan and Carl Sagan to Chuck Berry (1986) from https://www.loc.gov/item/cosmos000081, Used with permission.

LETTER 8 from 'A Cosmonaut's Letter: Gagarin's Letter to his Family, April 10, 1961' from Slava Gerovitch from *Soviet Space Mythologies: Public Images, Private Memories, and the Making of a Cultural Identity* from by Slava Gerovitch, © 2015. Reprinted by permission of the University of Pittsburgh Press.

LETTER 9 from Soviet cosmonauts to Leonid Brezhnev (1965) from Yuri Gagarin et al., 'Soviet Cosmonauts' Letter to Leonid Brezhnev, October 22, 1965,' in Slava Gerovitch, 'Computing in the Soviet Space Program,' http://web.mit.edu/slava/space/documents/brezhnev-letter.htm. Translated from the Russian by Slava Gerovitch. Reproduced by permission of the translator. The Russian original was published in Nikolai Kamanin, *Skrytyi kosmos*, vol. 2 (Moscow: Infortekst, 1997), pp. 245-248.

LETTER 10 from Roger Boisjoly to R. K. Lund (1985) from https://catalog.archives.gov/id/596263, courtesy NASA;

LETTER 13 from Alan Shepard to his parents (1959) from https://www.icollector.com/Alan-Shepard_i11118098, Used with permission.

LETTER 14 from Nellie Copeland to Dr William R. Kubinec (1985) from The Halley's Comet Project, 1985-1986: https://pascal-cofc.primo.exlibrisgroup.com/permalink/01PASCAL_COFC/frhrde/alma991009243859705613 Courtesy of Special Collections, College of Charleston Libraries, all rights reserved.

LETTER 16 from Geraldyn 'Jerrie' Cobb to US President John F. Kennedy (1963) from https://www.jfklibrary.org/asset-viewer/archives/JFKWHCNF/0515/JFKWHCNF-0515-002. Used with permission of the Estate of Jerrie Cobb.

LETTER 17 from Valentina Vladimirovna Zorkina to Valentina Vladimirovna Tereshkova. (1963) from Roshanna P. Sylvester, 'You Are Our Pride and Our Glory!' Emotions, Generation, and the Legacy of Revolution in Women's Letters to Valentina Tereshkova in *The Russian Review*, vol. 78, Iss 3, July 2019, with permission from Wiley via the Copyright Clearance Center.

LETTER 21 from Stanley Kubrick to Arthur C. Clarke (1964) from https://twitter.com/StanleyKubrick/status/9800518000828112896, Used with permission from Christiane Kubrick and Stanley Kubrick Film Archives LCC.

LETTER 22 dated May 1962 from Marion Carpenter to Malcolm Carpenter from 'Letter dated May 1962 from Marion Carpenter to Malcolm Carpenter' *For Spacious Skies: The Uncommon Journey Of A Mercury Astronaut* by Scott Carpenter and Kris Stoever. Copyright © 2002 by Scott Carpenter and Kristen C. Stoever. Reprinted by permission of Houghton Mifflin Harcourt Publishing Company. All rights reserved.

LETTER 24 from Julian Scheer to George M. Low (1969) from https://historical.ha.com/itm/explorers/first-words-spoken-on-the-moon-neil-armstrong-s-personal-copy-of-an-internal-nasa-document-regarding-whether-to-coach-him-on/a/6209-50058.s, Courtesy NASA.

LETTER 26 from Ray Bradbury to Arthur Schlesinger Jr and from John F. Kennedy to Ray Bradbury (1962) from https://jfk.blogs.archives.gov/2017/08/22/the-most-interesting-writer-about-the-future/, Ray Bradbury letter, courtesy of the Bradbury Estate, Used with permission from Don Congdon Associates and John F. Kennedy letter to Ray Bradbury, June 21 1962. White House Central Subject Files, Box 275, GI 2-8/B Literature, Books-Poetry (Executive).

LETTER 27 from Cara Lucy O'Connor to NASA, from https://www.irishtimes.com/news/ireland/irish-news/irish-schoolgirl-6-demands-nasa-make-pluto-great-again-1.3391792, Used with permission from Cara Lucy O'Connor.

LETTER 28 from William Safire to H. R. Haldeman (1969) from https://catalog.archives.gov/id/6922351, Series: H. R. Haldeman's Files, 1/20/1969 - 4/30/1973 Collection: White House Staff Member and Office Files (Nixon Administration), 1/20/1969–8/9/1974.

LETTER 29 from Neil Armstrong to James Whitman (2005) from Neil A. Armstrong Papers, Purdue University Archives and Special Collections, Purdue University Libraries, Used with permission.

LETTER 30 from Neil deGrasse Tyson to NASA (2008) from https://www.facebook.com/notes/neil-degrasse-tyson/an-open-letter-to-nasa/10157454721411613/, Used with permission from Dunow, Carlson & Lerner Literary Agency, Inc.

ACKNOWLEDGEMENTS

It requires a dedicated team of incredibly patient people to bring the *Letters of Note* books to life, and this page serves as a heartfelt thank you to every single one of them, beginning with my wife, Karina – not just for her emotional support during such stressful times, but for the vital role she has played as Permissions Editor on many of the books in this series. Special mention, also, to my excellent editor at Canongate Books, Hannah Knowles, who has somehow managed to stay focused despite the problems I have continued to throw her way.

Equally sincere thanks to all of the following: Teddy Angert and Jake Liebers, whose research skills have helped make these volumes as strong as they are; Rachel Thorne and Sasmita Sinha for their crucial work on the permissions front; the one and only Jamie Byng, whose vision and enthusiasm for this series has proven invaluable; all at Canongate Books, including but not limited to Rafi Romaya, Kate Gibb, Vicki Rutherford and Leila Cruickshank; my dear family at Letters Live: Jamie, Adam Ackland, Benedict Cumberbatch, Aimie Sullivan, Amelia Richards, and Nick Allott; my agent, Caroline Michel, and everyone else at Peters, Fraser & Dunlop; the many illustrators who have worked on the beautiful covers in this series; the talented performers who have lent their stunning voices not just to Letters Live, but also to the *Letters of Note* audiobooks; Patti Pirooz; every single archivist and librarian in the world; everyone at Unbound; the team at the Wylie Agency for their assistance and understanding; my foreign publishers for their continued support; and, crucially, my family, for putting up with me during this process.

Finally, and most importantly, thank you to all of the letter writers whose words feature in these books.

ACKNOWLEDGMENTS